Deep Signal: AI and IoT Unlocking Underground Mining Communication

GoodMan, Volume 1

Patrick Mukosha

Published by Patrick Mukosha, 2025.

Deep Signal: AI and IoT Unlocking Underground Mining Communication

"Transforming Mines, Connecting Minds: Intelligence, Efficiency, and Safety Below the Surface."

Copyright Notice

First Published: September, 2025

Publisher: Dr. Patrick Mukosha

Trademarks

Warning and Disclaimer

Author: Dr. Patrick Mukosha

Abstract

Underground mining has remained a difficult communication and connectivity frontier in a time when digital transformation affects every business. Underground operations are particularly challenging due to harsh conditions, intricate topologies, and crucial safety requirements.

Deep Signal examines how the Internet of Things (IoT) and artificial intelligence (AI) are not only instruments but also ground-breaking facilitators of dependable, fast speech and data transmission in mining settings.

For mining executives, technology strategists, engineers, and researchers, this book provides a forward-looking perspective on the intersection of artificial intelligence, the Internet of Things, and underground communication.

Deep Signal demonstrates how intelligent connectivity can revolutionize productivity, safety, and profitability through everything from real-time hazard alerts and dynamic network optimization to predictive maintenance and autonomous vehicle coordination.

- Patrick Mukosha

The Significance of This Book

Deep Signal serves as a link between the harsh reality of deep mining and new technologies. It gives experts a road plan for converting their operations into secure, lucrative, and technologically advanced systems. This is a forward-thinking guidebook for the next phase of mining communication, not just a technical one.

Chapter 1: The Underground Challenge: Why Conventional Communication Is Ineffective

1.1. Deep Signal Challenges

One of the most difficult and complicated situations for communication systems is underground mining. Due to special geological, structural, and operational considerations, *underground mines interfere with traditional communication techniques in contrast to surface industry where satellite links, Wi-Fi, and mobile networks operate without a hitch.*

Instantaneous communication is essential to the world above ground. Connectivity promotes productivity, security, and profitability in everything from 5G networks that power driverless cars to real-time data analytics in smart factories.

However, *mining activities face difficulties in a digital desert beneath the surface of the Earth.* Underground mines are harsh, complicated places. Tunnels have unpredictable twists and turns. Signals are absorbed by rock strata. Waves are reflected and scattered by metallic structures. Communication is further hampered by dust, moisture, and vibrations. Miners have been using costly, rigid, and failure-prone wired systems, leaky feeder networks, or low-frequency radio for decades. Every communication outage has the potential to cause lost productivity, downtime, and—most importantly—life-threatening risks.

1.1.1. Signal Attenuation and Absorption

- **Rock and Soil Interference:** *Signals from radio, Wi-Fi, and cell phones become significantly weaker when they pass through thick geological layers.* Electromagnetic waves are absorbed or scattered by barriers made of rocks, minerals, and moisture.
- **Depth Factor:** Signal strength and dependability significantly decrease as the depth of the mine increases due to the longer distance that signals must travel.

1.1.2. Reflection and Multipath Propagation

- *Signals are reflected and diffracted by the uneven geometries created by tunnels, shafts, and stopes.*
- This results in multipath interference, which causes distortion, delays, and erratic connections when signals reach receivers out of phase.

1.1.3. Dynamic Mining Environment:

- As layouts are altered by blasting, drilling, and tunnelling, mines are always changing.
- *When routes vary, conventional technologies like wired communication or fibre-optic cables are susceptible to damage or obsolescence.*
- Why Re-establishing wired networks after frequent moves is expensive and time-consuming.

1.1.4. Electromagnetic Interference (EMI):

- *Wireless signals are interfered with by electromagnetic noise produced by electrical systems, heavy machinery, and drilling equipment.*
- In these high-noise areas, traditional communication systems made for clean settings are unable to remain stable.

1.1.5. Limited Line-of-Sight:

- *Line-of-sight is necessary for many conventional communication methods, such as microwave lines.*
- Subterranean tunnels divide, bend, and curve, removing straight signal routes.

1.1.6. **Environmental and Safety Restrictions**

- Dust, excessive humidity, and explosive gasses are common in underground mines. *Conventional electronics soon deteriorate if they are not made to survive these extreme circumstances.*
- Longevity and safety issues with battery-powered devices arise in volatile or oxygen-deficient situations.

1.1.7. **Bandwidth Legacy Systems' Drawbacks**

- As mining moves toward digital transformation, traditional methods cannot support real-time, data-driven operations;
- Older systems, such as leaky feeder radios, despite their widespread use, have limited capacity for data-heavy applications like video surveillance, IoT sensor feeds, and autonomous equipment control.

1.1.8. **Emergencies Highlight Your Weaknesses**

- *Conventional wired systems are frequently destroyed and wireless communications are blocked in the event of rock falls, fires, or cave-ins.*
- This hinders rescue efforts and jeopardizes worker safety during crucial communication moments.

In summary, because it was never intended for such intricate, dynamic, and signal-hostile settings, conventional communication fails underground. Specialized AI and IoT-driven communication systems that can adapt, self-optimize, and maintain resilience even under the most extreme circumstances are required due to the geological makeup, safety regulations, and operating dynamics of mines.

1.2. The Cost of Disconnection

Think about it: *a single misunderstanding can result in hours of subterranean operational delays. Machine movements that are not coordinated run the risk of colliding. Failures of equipment go ignored. Personnel do not receive emergency alerts in a timely manner.* With the correct technology, the ongoing conflict between operational effectiveness and human safety in mines can be avoided and needless.

Signal loss is only one aspect of the problem; *complexity* is another. Drilling rigs move, tunnels widen, and ventilation systems change, making mines dynamic ecosystems. A network that is static soon becomes outdated. Conventional methods of subterranean communication are reactive rather than proactive, which limits their flexibility.

1.3. A Look at Intelligence in the Future: Below the Surface

The Internet of Things (IoT) and artificial intelligence (AI) promise a fundamental revolution rather than merely small-scale advancements. Imagine a mine where:

- AI algorithms anticipate *signal interference* and dynamically reroute data to maintain seamless voice and telemetry communication;
- Autonomous vehicles and human operators collaborate in real time, coordinated by cognitive networks that optimize every movement;
- Emergency alerts are not just sent—*they are predicted*, ensuring miners reach safe zones before hazards materialize; and
- Sensors embedded in walls, machinery, and miners continuously feed environmental and operational data to an intelligent system.

This isn't science fiction. A subterranean intelligence network that adapts, learns, and protects life and productivity is the frontier that the mining industry is just starting to explore.

1.4. The Reasons AI and IoT Will Revolutionize Mining

1.4.1. Dynamic Signal Management

Artificial intelligence (AI) algorithms are *transforming the way radio frequencies and bandwidth are managed in underground mines, guaranteeing smooth voice and data connection even in challenging, blocked situations.* Due to signal attenuation brought on by machinery, tunnels, and rock formations, traditional fixed communication systems frequently malfunction in mines.

- **How It Operates**: AI keeps an eye on the network in real time, rerouting signals and reallocating bandwidth to maximize connectivity. Machine learning algorithms anticipate signal degradation and make proactive adjustments to transmission power or communication channel switching.
- **Example: Sibanye Stillwater mine in South Africa** used AI-driven network optimization to keep its deep-level platinum miners' communications constant. As a result, there were fewer dropped calls and a 30% increase in data throughput.
- **Case Study:** Adaptive wireless networks for mining have been developed by companies such as **ABB and Cisco**, where AI continuously assesses environmental factors and signal quality to dynamically regulate frequency allotment. This makes it possible for real-time voice communications and high-speed telemetry, which are crucial for autonomous equipment coordination.

1.4.2. Hazard Identification And Predictive Maintenance

Large volumes of operating data, including temperature, vibration, pressure, and stress levels, are gathered by IoT sensors integrated into mining machinery. *Predictive analytics driven by AI analyse this data to anticipate malfunctions before they happen, averting expensive downtime and disastrous mishaps.*

- **How It Operates**: A centralized AI system receives continuous streams of performance measurements from sensors. Algorithms automatically initiate maintenance programs, detect irregularities, and spot trends connected to possible failure.
- **Example: Rio Tinto's Mine of the Future** initiative makes use of IoT-enabled drills and trucks that provide real-time mechanical health reports. AI anticipates maintenance requirements, averting malfunctions that can cause a days-long halt in operations.
- **Hazard Detection:** IoT also keeps an eye on environmental factors including water intrusion, earthquake activity, and gas levels. AI, for example, is able to identify minute changes in sensor data that occur before a rock fall or gas release.
- **Case Study:** By combining IoT sensors with predictive maintenance, the **Codelco Chuquicamata mine in Chile** has improved safety and operating efficiency by reducing machinery downtime by more than 20%.

1.4.3. Efficiency in Operations

Throughput is increased, *downtime is decreased, and resource allocation is improved* when real-time voice, telemetry, and data communications are integrated.

- **How It Operates**: AI optimizes scheduling and operations by processing data streams from many systems, including communication networks, ventilation systems, conveyor belts, and autonomous trucks. It facilitates quick decision-making by offering a centralized perspective of the mine.

- **Example: BHP's underground operations in Australia** use AI-driven scheduling systems to manage self-sufficient loaders, haul trucks, and drilling equipment, minimizing downtime and optimizing the effectiveness of ore extraction.
- **Case Study:** AI is able to dynamically balance workloads at **Vale's Carajás Mine in Brazil** thanks to IoT-enabled monitoring of conveyor systems and crushers, preventing bottlenecks and lowering energy consumption by up to 15%.

1.4.4. Designing for Safety

By evaluating sensor data from the mining environment to *foresee risks and automatically initiate preventive measures,* artificial intelligence (AI) improves safety. In a field where rock falls, floods, and gas leaks can be lethal, this proactive approach is essential.

- **How It Works:** To find early warning indicators, AI models examine patterns in sensor readings, such as water pressure, structural strain, or methane levels. Control rooms, miners' wearable technology, or automated safety systems like barriers or ventilation fans can all get alerts.
- **Example: Glencore's Mount Isa Mines in Australia** employ AI to track subterranean conditions, automatically turning off machinery and turning on ventilation when gas concentrations rise above acceptable limits.
- **Case Study**: Over the course of three years, a **Chinese coal mine** reduced explosion incidences by 40% by integrating IoT sensors with AI-based gas detection. By simulating possible

risk scenarios and analysing real-time data, the technology enables proactive evacuation or mitigation.

In summary, when AI and IoT are combined, subterranean mining becomes proactive rather than reactive. In the end, *they boost productivity, cut expenses, and save lives by offering dynamic communication, predictive maintenance, operational optimization, and improved safety.* **Rio Tinto, Vale, BHP**, and **Codelco** are just a few of the *mines that have shown quantifiable increases in productivity, safety, and profitability since implementing these technology.*

1.5. The Path Ahead

The underground problem is strategic rather than merely technical. Mines that don't adjust will have to deal with growing expenses, more risk, and a decline in their competitiveness. Faster, safer, more effective, and more robust underground communication will become a strategic advantage for those who adopt AI and IoT. *This book will walk you through the strategies, technologies, and practical applications that enable this change.* **Deep Signal** lays out the roadmap for the future hyper-connected mine, from AI-driven signal optimization to dynamic sensor networks, from autonomous machine coordination to predictive hazard management.

This is where the adventure starts, underground, where industry and intelligence collide and communication becomes essential to underground mining.

Chapter 2: Digital Twin Mines: Charting the AI-Powered Understanding

Overview

A *virtual, data-rich, and dynamic depiction of a physical asset, procedure, or system that replicates its real-world counterpart across the course of its existence* is called a **Digital Twin**. It simulates, analyses, and optimizes the physical object's performance using real-time data from sensors, allowing for better decision-making, predictive maintenance, and scenario testing prior to practical implementation.

The **Digital Twin Mine**, *a live, intelligent model of underground operations that transforms data into actionable information*, is the answer to the problems with underground communication that were described in Chapter 1. A 3D map is only one aspect of a digital twin. With the integration of geological data, machine telemetry, ambient sensors, and human activity, it is a dynamic, real-time virtual representation of the mine. This twin uses AI and IoT to forecast, optimize, and protect reality rather than merely reflecting it.

2.1. A Digital Twin Mine's Anatomy

2.1.1. Geospatial Mapping

Accurate, constantly updated geospatial data is the foundation of a digital twin mine. Tunnels, shafts, stopes, and ore bodies can all be accurately depicted at the centimetre level using high-resolution 3D mapping. Technologies like photogrammetry, ground-penetrating radar, and LIDAR (Light Detection and Ranging) produce intricate models that capture the constantly shifting environment of the mine.

- **Real-Time Updates**: AI algorithms identify changes, deviations, or subsidence during blasting or excavation, notifying engineers before structural issues become serious.
- **Example:** Rio Tinto's "Mine of the Future" project tracks alterations in ore bodies and tunnel integrity using 3D mapping and AI, enhancing operational safety and resource estimation.
- **Case study:** At **BHP's Olympic Dam**, engineers were able to identify early indications of rock movement through continuous photogrammetry and LIDAR scans of underground tunnels, averting possible collapses and improving support structures.

2.1.2. Integration of Equipment

All of the working equipment in modern mines is integrated into the digital twin, making them intelligent devices. IoT sensors installed on drills, loaders, haul trucks, and conveyor belts send data regarding location, stress levels, and performance continuously.

- **Predictive Maintenance:** AI systems examine vibration, temperature, and motor current trends to anticipate problems before they arise. This prolongs the life of the equipment and lowers maintenance expenses and downtime.
- **Operational Optimization**: Workload balancing, bottleneck reduction, and fuel efficiency can all be achieved by dynamically adjusting scheduling.
- **Example: Goldcorp's Borden Mine** in **Canada** installed an autonomous haulage fleet that was incorporated into its digital twin, which increased overall efficiency and cut down on truck idle time by 20%.
- **Case Study**: The digital twin at **Anglo American's Mogalakwena Mine** can simulate various equipment configurations, forecasting throughput changes and streamlining extraction procedures thanks to integrated machine data.

2.1.3. Monitoring of the Environment

In underground mining, sustainability and safety are crucial. Temperature, humidity, dust, vibrations, and gas concentrations (such as carbon monoxide and methane) are all continuously monitored by Internet of Things sensors.

- **Predictive Hazard Detection**: AI algorithms analyse these streams to foresee potentially hazardous situations, such the accumulation of toxic gas or the exploding of rocks.
- **Optimisation of mine Conditions**: Mine conditions can be optimized by dynamically adjusting lighting, ventilation systems, and emergency procedures in response to real-time

data.

- **Example: Vale's Salobo Mine** in **Brazil** employs an Internet of Things sensor network to monitor gas and dust; data entered into a digital twin allows for predictive responses, lowering safety and environmental hazards.
- **Case study:** By using a digital twin for environmental monitoring, **Newmont's Boddington Gold Mine** was able to optimize ventilation using AI, which resulted in a 15% reduction in energy usage while preserving safe subterranean air quality.

2.1.4. Human Communication

Miners themselves are included in digital twins, in addition to machinery. Miners' position, vital signs, and level of weariness are tracked in real time using wearable technology, such as smart helmets, vests, and wristbands.

- **Predictive Safety Alerts**: Safety officers receive notifications if a miner's heart rate or degree of weariness reaches dangerously high levels, and workload or rotation schedules can be automatically adjusted.
- **Emergency Coordination:** The digital twin streamlines evacuation and medical operations in the event of an accident by giving rescue teams real-time location and health status information.
- **Example: Kumba Iron Ore in South Africa** used wearable sensors connected to its digital twin, enabling predictive notifications for weariness and early identification of heat stress occurrences.
- **Case Study**: **Newcrest Mining** tested wearable biometric sensors underground in Australia; by combining this data with the digital twin, worker health could be predicted and

emergency response times could be shortened.

2.1.5. Nodes of Communication

A strong communication network serves as the foundation of a digital twin mine. Every wearable, machine, and sensor turns into a node that sends data in real time.

- **Optimization of AI**: Even under difficult subterranean conditions, the system continuously optimizes data paths to minimize latency and guarantee continuous voice, telemetry, and video signal delivery.
- **Resilience:** AI reroutes data to provide operational continuity and safety monitoring in the event of partial network failures.
- **Example: Sandvik's AutoMine®** technology uses a mesh network to link wearables, sensors, and autonomous machines that are incorporated into the digital twin for real-time operational control.
- **Case Study**: A multi-node communication network at **Boliden's Garpenberg Mine** guarantees constant telemetry and video feeds from subterranean cameras to the control centre, enabling prompt action in the event that irregularities are noticed.

In summary, the digital twin mine is a dynamic, living model that integrates intelligent communication, human integration, environmental monitoring, intelligent machinery, and exact mapping. When combined, these layers enable predictive decision-making that can save lives, cut expenses, and maximize resource extraction while also enabling safer, more effective, and sustainable mine operations.

2.2. The Transformational Potential of Digital Twins

2.2.1. Making Proactive Decisions

Mining operations can transition from reactive to proactive management with the help of digital twins. *Before making adjustments in the real world, operators can practice various operational scenarios by building a virtual version of the mine.* These simulations are analysed by AI algorithms to forecast possible equipment failures, bottlenecks, or safety risks.

- **Example:** A digital twin was used to model the effects of various drill-and-blast schedules in an **Australian gold mine**. During periods of peak output, the AI forecasted a significant probability of congestion on haulage routes. The mine decreased operational risks and prevented expensive delays by modifying the drilling timetable in the virtual twin.
- **Case Study:** Digital twins are used by Rio Tinto's iron ore facilities in Western Australia to predict equipment failures. Predictive maintenance is made possible by sensors that provide the twin with real-time data. According to reports, this strategy decreased unscheduled downtime by up to 20%, resulting in yearly cost savings of millions.

2.2.2. Optimization of Operations

Digital twins serve as a mechanism for making decisions in real time. They keep an eye on and evaluate human movement, equipment efficiency, and resource allocation. Mines can increase output, minimize waste, and optimize operations by identifying inefficiencies.

- **Example:** A **Chilean Copper Mine** optimizes truck dispatch timetables and conveyor belt speeds using its digital counterpart. Without adding more workers or equipment, the system found that changing loader assignments might increase overall productivity by up to 15% of operating time every shift.
- **Case Study**: Digital twins are used to replicate ore extraction processes at **BHP's operations.** Managers were able to improve extraction efficiency and energy usage at the same time by real-time redistributing personnel and equipment thanks to the twin's observations.

2.2.3. Guaranteed Safety

In mining, *worker safety is a top priority, and digital twins greatly improve hazard control.* The twin can identify dangerous situations like gas leaks, rock falls, or equipment failures by combining IoT sensors with artificial intelligence. It can then send miners predictive notifications before disasters happen.

- **Example:** A digital twin keeps an eye on machine vibrations, seismic activity, and subterranean air quality in a platinum mine in South Africa. The system allows for proactive maintenance and evacuation by warning supervisors of possible collapses or high-risk areas.
- **Case Study:** To model evacuation pathways in the event of an emergency, **Kirkland Lake Gold** deployed a digital twin. They greatly decreased the possibility of injuries by optimizing escape routes and emergency response procedures through the analysis of several situations.

2.2.4. Continuous Learning

Digital twins are dynamic models that change as the mine does. AI models are constantly learning from operational data, adjusting to shifts in the environment, equipment use, and mine architecture. Over time, the system becomes more resilient and smarter as a result of this ongoing improvement.

- **Example:** A digital twin that is updated in response to continuous production measures was installed at a **Lithium Mine in Canada**. The mine can better foresee maintenance requirements and supply limitations thanks to AI algorithms that recalculate predictions for equipment wear and resource consumption every week.
- **Case Study:** To improve prediction algorithms, **Anglo American's Mogalakwena Platinum Mine** uses a digital twin that combines historical and current data. By learning from every shift, the twin has gradually increased operational efficiency and safety results, thus becoming into a "smarter" mine with each repetition.

In conclusion, digital twins are revolutionary because they combine continuous learning, predictive analytics, and real-time monitoring into a single, integrated system. They enable mining businesses to take preventative measures, optimize operations, protect workers, and adapt to the mine's changing needs, resulting in quantifiable improvements in productivity, profitability, and sustainability.

2.3. The Way Ahead: Cognitive Underground Mines

Digital twins are just the first step. Mines can transition from reactive to cognitive subsurface operations by integrating AI and IoT. These mines will:

2.3.1. How Mining Operations Are Transformed by Advanced Digital Twins

2.3.1.1. Use Real-Time and Historical Data to Self-Optimize Operations

Digital twins connect previous operational data with live sensor inputs, producing a feedback loop that continuously optimizes mine performance. Artificial intelligence (AI) algorithms can recommend changes to increase productivity, cut costs, and conserve energy by examining trends from previous production cycles, equipment performance, and environmental factors.

- **Example:** The digital twin of a **Canadian Nickel Mine** receives historical data from drills, loaders, and haul trucks. The system found that, without the need for further equipment, bottlenecks could be reduced by 12% by changing shift patterns and conveyor speeds.
- **Case Study:** Digital twins are used to examine years' worth of production and maintenance data from **Vale's Iron Ore Plants in Brazil.** In order to maximize flow rates and energy consumption, the system automatically modifies the operational parameters for transport vehicles, conveyors, and crushers, resulting in quantifiable productivity benefits.

2.3.1.2. Autonomously Organize Staff and Equipment for Optimal Performance

For mining operations, digital twins serve as the main source of intelligence. The twin can optimize route planning, assign resources dynamically, and coordinate tasks by continuously monitoring both personnel and equipment. This coordination reduces downtime, prevents disputes, and boosts output.

- **Example:** To dynamically arrange ore pickups and deliveries at a **Copper Mine in Chile**, self-governing loaders and trucks interact with the digital twin. Human operators are given instructions on effective and safe paths in the interim. This autonomous coordination increased total mine throughput and cut down on truck idle times by 18%.
- **Case Study: Rio Tinto's "Mine of the Future**" initiative serves as a large-scale example of this. The digital twin directs autonomous haul trucks and drilling rigs, prioritizing machine utilization, managing work sequencing, and guaranteeing the safe deployment of humans. As a result, operational efficiency is greatly increased while human exposure to dangerous situations is decreased.

2.3.1.3. Foresee and Avoid Risks Before They Appear

Adoption of digital twins is *significantly influenced by safety*. In order to anticipate dangers before they materialize, artificial intelligence (AI) in the twin continuously evaluates sensor inputs, such as temperature variations, gas levels, equipment vibration, and seismic activity. Early warnings enable actions to stop mishaps, equipment malfunctions, or environmental problems.

- **Example:** A digital twin is used by a platinum mine in South Africa to track seismic micro-events. By anticipating possible rock falls and automatically setting off alarms, the device makes it possible to fortify susceptible tunnels and conduct preventative evacuations.
- **Case Study:** The digital twin at **Anglo American's Mogalakwena mine** combines ventilation readings, geotechnical data, and vibration sensors. Early warning indicators of possible stope collapse were successfully identified by the predictive model, enabling engineers to fortify the region and avert severe downtime or injuries.

2.4. Constantly Change as New Data Takes on Different Shapes Knowledge of Underground Dynamics

Living systems are digital twins. The model gains fresh information from every shift, machine activity, and environmental change, which helps it better comprehend the mine. Predictive accuracy, operational effectiveness, and safety are all improved over time by this ongoing learning.

- **Example:** a **Lithium Mine in Australia** updates its digital twin every day with information on ventilation, equipment operation, and ore quality. AI systems constantly increase efficiency and dependability by recalibrating forecasts for extraction processes and maintenance plans.
- **Case Study:** Digital twins that adjust to shifting geological conditions are used in **BHP's underground operations**. The technology detects minute changes in ore body geometry and rock stability by integrating real-time data. Predictive models are updated by the AI in accordance with these changes, improving operational choices and decreasing downtime.

All of these features put digital twins in a position to function as independent operational command centres. They allow mines to proactively prevent dangers, seamlessly coordinate humans and robots, self-optimize, and continuously adapt in response to new information. A safer, more effective, and more intelligent mining operation that can adapt quickly to the intricacies of subterranean surroundings is the end result.

With every sensor, machine, and human interconnected in a real-time, adaptive ecosystem, the Digital Twin Mine signifies a paradigm shift in mining from extraction to intelligent orchestration.

From IoT sensor networks to AI-driven signal intelligence, we will delve deeper into the technology behind these advancements in the upcoming chapters. We will also examine how these systems come together to form the future's hyper-connected, ultra-efficient mines.

Chapter 3: Surface Networks: Actuators, Sensors, and the Connectivity Backbone

Overview

The *Internet of Things network* is a mine's *nervous system*, if its *digital twin* is its *brain*. It is impossible to make decisions in real time without a strong, intelligent network that links sensors, equipment, and people. This chapter examines how IoT facilitates smooth communication underground, serving as the foundation for autonomous operations, AI-driven optimization, and predictive analytics. Connectivity between miners, equipment, and the physical world has always been essential to mining.

This connectivity is no longer restricted to wires, radios, or human messengers thanks to the integration of the *Internet of Things* (IoT). Rather, underground mines are evolving into living, breathing ecosystems where sensors and actuators work together to create an intelligent neural network that constantly observes, reacts, and adjusts to the ever-changing subsurface environment.

This chapter examines how the digital nervous system of contemporary underground activities is made up of IoT devices and connectivity backbones.

3.1. The Function of Sensors: The Mine's Eyes and Ears

3.1.1. The Role of Sensors: Eyes and Ears of the Mine

The perception layer of subterranean IoT systems is made up of sensors. They act as a linked mine's eyes, ears, and touch points.

- **Environmental Monitoring**:
    - Gas sensors immediately notify control systems when thresholds for methane, carbon monoxide, and other toxic gases are reached.
    - Temperature and humidity sensors monitor changes in the microclimate, preventing condensation that could harm machinery and guaranteeing that worker air circulation is safe.
    - By detecting early-warning tremors that are imperceptible to the human senses, seismic and vibration sensors predict rock falls or structural instability.
- **Operational Monitoring:**
    - Conveyor belts, loaders, and drills all include pressure and load sensors that gauge stress levels in order to forecast failures.
    - By listening to the sounds of equipment, acoustic sensors can identify irregularities like misalignments or bearing wear.
    - By detecting human activity close to self-driving cars or large machinery, proximity sensors improve safety.
- **Case Study: Rio Tinto's "Mine of the Future"**

Thousands of Internet of Things sensors are used by Rio Tinto's Pilbara iron ore facilities to track drilling precision, conveyor belt tension, and truck payloads. By improving predictive maintenance and increasing equipment uptime, this level of information has decreased expensive shutdowns.

3.1.2. **Actuators: The System's Hands and Muscles**

Actuators carry out decisions if sensors gather information. These gadgets *transform intelligence into movement by translating computer commands into tangible actions.*

- **Ventilation Systems**: In response to sensor data, automated fans and dampers move airflow where it is most needed. AI-powered on-demand ventilation can reduce energy use by as much as 40%.
- **Water Management**: Actuators regulate pumps that prevent subterranean flooding and make sure water doesn't damage equipment or tunnels.
- **Robotics and Autonomous Vehicles**: Actuators allow drilling equipment, robotic arms, and self-driving loaders to function precisely under challenging circumstances.
- **Safety Systems:** When dangerous thresholds are crossed, actuators activate alarm systems, fire suppression sprinklers, and emergency barriers.
- **Example:** IoT sensors and actuator-driven ventilation systems have been used in **South African Gold Mines** to autonomously modify air quality, greatly lowering worker exposure to dangerous chemicals.

3.1.3. **The Foundation of Communication: Dark Networks**

In the absence of a dependable communication backbone, all of these gadgets are useless. Underground mining has distinct difficulties compared to urban IoT systems: radio signals are absorbed by rock layers, wave propagation is distorted by tunnels, and equipment frequently produces interference.

- **Wireless Communication Options**:
    - **Leaky Feeder Systems**: Coaxial cables that purposefully leak signals enable constant two-way data and radio transmission over tunnels.
    - **Wi-Fi Mesh Networks**: Self-healing networks are produced by ruggedized routers, which guarantee that data finds alternate routes in the event that a node fails.
    - **5G Underground**: New private 5G networks offer real-time machine-to-machine communication with extremely low latency and great bandwidth.
    - **Acoustic and Optical Links:** Li-Fi (light-based communication) and acoustic (ultrasonic) transmissions are being investigated for specialized uses in areas where radio waves are ineffective.
- **Integration of Edge and Fog Computing:** Data processing frequently takes place nearer the source to minimize latency. Before transmitting data to the surface, edge devices locally combine sensor readings, filter out noise, and rank the most important alarms.

3.1.4. **Case Study: Sweden's Boliden's Kankberg Mine**

Boliden collaborated with ABB and Ericsson to construct a 5G-enabled subterranean network. The system runs on a secure backbone and supports augmented reality headgear, remote drilling, and driverless cars. Better worker safety and a notable drop in operating expenses were the outcomes.

3.1.5. Underground Energy Powering The Internet Of Things

This matter is of great importance. IoT devices must operate constantly in remote, challenging-to-reach locations.

- **Battery-Powered Sensors:** Years of battery life are extended by low-energy protocols like NB-IoT and LoRaWAN.
- **Energy Harvesting:** Advancements like thermal gradient harvesters and vibration-powered sensors do away with the necessity for regular battery replacement.
- **Smart Power Management:** To save energy, systems turn off non-essential equipment when they are not in use.

3.1.6. Difficulties and Prospects

- **Scalability:** There must be no congestion when thousands of devices join.
- **Durability:** To survive vibration, heat, moisture, and dust, sensors and actuators must be ruggedized.
- **Cybersecurity:** All linked nodes have the potential to be targets. Network segmentation and robust encryption are crucial.
- **Interoperability:** Legacy equipment is frequently used in mines. It's still difficult to guarantee smooth connection between the new and old systems.
- **Future Vision:** Completely autonomous "*Cognitive Mines*" in which Internet of Things devices create networks that organize themselves and dynamically change communication priorities, power consumption, and routes without the need

for human interaction.

3.1.7. **Key Takeaway:** The digital mine's nervous system, which combines remote machines and human operators into a unified, intelligent environment, is made up of sensors, actuators, and subterranean communication backbones. The Internet of Things network is a mine's nervous system, if its digital twin is its brain. *It is impossible to make decisions in real time without a strong, intelligent network that links sensors, equipment, and people.* This chapter examines how IoT facilitates smooth communication underground, serving as the foundation for autonomous operations, AI-driven optimization, and predictive analytics.

3.2. Comprehending Underground IoT

In mining, the *Internet of Things* (IoT) is the distributed intelligence network that supports the digital mine, not just a dispersed collection of devices buried in shafts and tunnels. *Every sensor, actuator, and wearable works as a node in a living neural web rather than as a standalone part, sending and receiving data, carrying out AI-driven commands, and preserving situational awareness in real time.*

An important advancement is underground IoT, which turns mines into self-aware, adaptable systems that continuously optimize production efficiency, worker safety, machine performance, and environmental monitoring. However, there will be challenges in realizing this objective.

Underground networks must function in one of the harshest and least predictable environments on Earth, in contrast to urban or industrial surface areas.

3.2.1. **Signal Attenuation: The Quiet Threat to Communication**

- **The Challenge**: Radio waves, particularly those at higher frequencies, are absorbed and scattered by rock strata, significantly decreasing signal power and range. Tunnels with sharp bends produce "shadow zones," and interference is exacerbated by mineral content like iron ore or sulphides.
- **Impact:** Unreliable connectivity can interfere with worker-equipment collaboration, delay important notifications, and jeopardize autonomous machine navigation.
- **Remedies:**
    - Leaky feeder systems use coaxial cables that function as "artificial antennas" to send radio frequency communications down tunnels.

  - Devices can relay data across several hops on mesh networks, avoiding restricted zones.
  - Redundancy is ensured via hybrid systems that combine RF, Wi-Fi, Li-Fi (light-based communication), and even ultrasonic communications.

3.2.2. Tough Environments: The Toughest Devices Survive

- **The Challenge**: Electronics are put to the test under the high-dust, high-vibration, and high-humidity conditions found in mines. Wear and failure are accelerated by corrosive gasses, water seepage, and temperature changes.
- **Impact:** Regular device failures raise maintenance expenses, erode confidence in IoT systems, and put safety-critical areas at risk for data blind spots.
- **Remedies:**
  - Ruggedized enclosures (IP67/68-rated casings) offer protection from water and dust.
  - Circuit boards are protected from corrosion and moisture by formal coatings.
  - Vibration-induced failures are decreased by shock-resistant mounting.
  - The network self-monitors its own health thanks to predictive maintenance sensors on the IoT devices themselves.

3.2.3. A Network That Has to Change With the Mine: Dynamic Topology

- **The Problem:** Underground mines are dynamic, in contrast to manufacturing. As excavation continues, tunnels may widen, contract, or change. Every modification modifies the communication paths' geometry.

- **Impact:** Networks run the danger of fragmenting, making freshly established workspaces unusable while decommissioned portions continue to use resources.
- **Remedies:**
    - When a node is lost, self-healing mesh protocols automatically reroute data.
    - AI-powered topology management that anticipates tunnel extensions and reallocates resources in advance.
    - Plug-and-play IoT nodes, or modular network kits, which can be quickly set up in new locations.
- **Example: LKAB's Kiruna Mine in Sweden**: AI-assisted mesh networks reduced downtime and eliminated the need for manual rewiring when portions of the mine were extended by automatically reconfiguring connectivity.

3.2.4. **Connectivity Without Compromise: Safety Restrictions**

- **The Problem:** Explosive gasses like methane are found in mines. In order to avoid sparks, overheating, or energy discharges that could ignite combustible atmospheres, electronic systems must be inherently safe.
- **Impact:** Risk tolerance is almost zero since any failure to comply with safety regulations could result in devastating explosions.
- **Remedies:**
    - Limiting voltage, current, and heat in IoT devices using intrinsic safety (IS) design.
    - **Wireless substitutes** to lessen the need for high-voltage wiring.

  - **Redundant Safety Procedures**: equipment must shut down gracefully rather than unexpectedly in the event of a fault.
  - **Constant AI monitoring**: making sure that IoT nodes in dangerous areas never go over operational limits.

3.2.5. AI-Enabled Protocols and Hybrid Architectures for a Resilient IoT

Hybrid communication architectures, which include several overlapping technologies so that others may easily take over when one fails, are the key to the future of underground IoT. As an illustration, consider 5G and Wi-Fi networks for high-bandwidth machine communication.

- Voice and emergency signalling via a leaky feeder radio frequency.
- Line-of-sight tunnel monitoring via Li-Fi.
- Long-lasting sensors in far shafts using low-power LoRaWAN.

Adaptive protocols powered by AI further improve resilience:

- **Dynamic Spectrum Allocation**: To prevent interference, AI modifies frequencies in real time.
- **Load balancing:** to avoid bottlenecks, data flow is intelligently spread.
- **Predictive routing:** AI reroutes data via the best routes by anticipating weak zones before they materialize.

3.2.6. **Key Takeaway**: The goal of underground IoT is to *create an autonomous nervous system that can function independently of geological interference, severe conditions,* *changing tunnels, and stringent safety regulations*. This goes beyond simply setting up sensors and radios. Subterranean IoT becomes the intelligent mine's lifeline by fusing hybrid architectures with AI-enabled adaptability, guaranteeing real-time communication, efficiency, and safety.

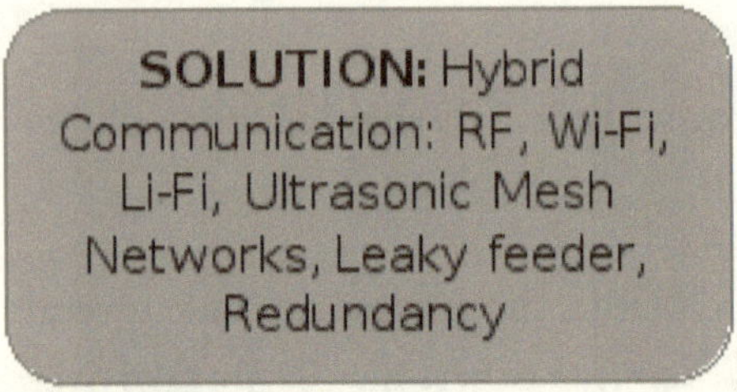

Diagram: Underground IoT Challenges & Solutions

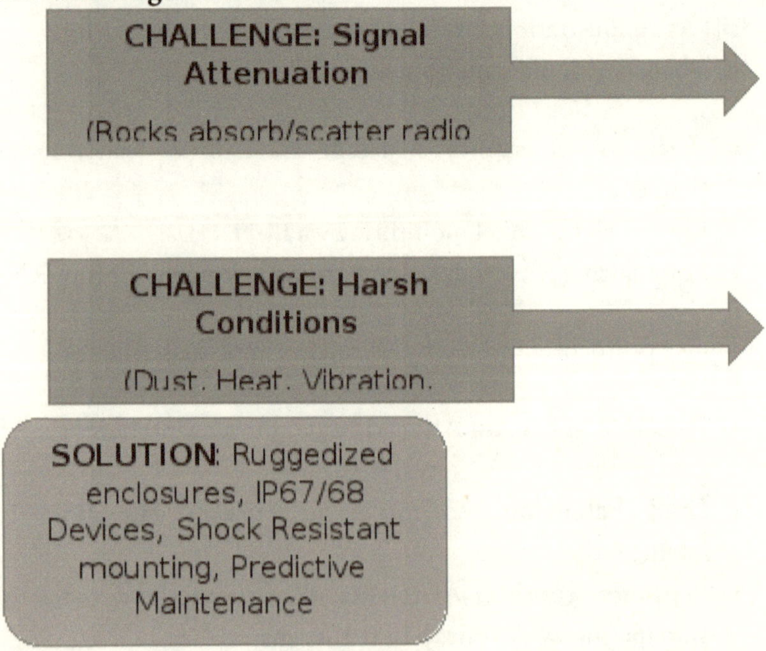

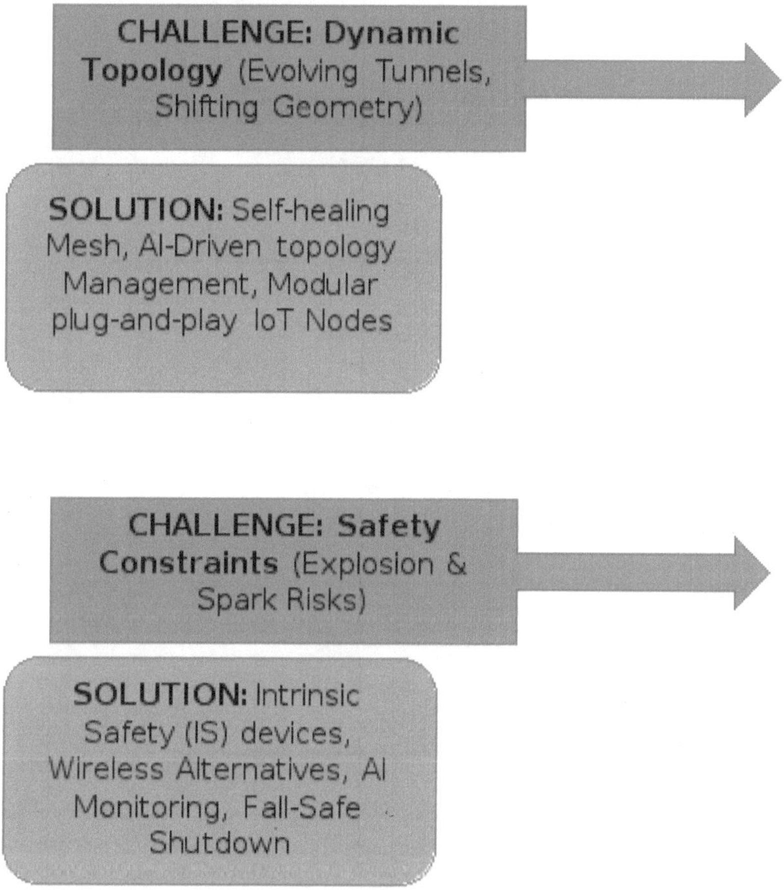

Figure 1: Underground IoT Challenges & Solutions

This is the map of Underground IoT Challenges & Solutions. It makes the section visually intuitive for your readers by matching each distinct underground IoT difficulty with the matching set of answers.

Diagram: Underground IoT Layered Architecture

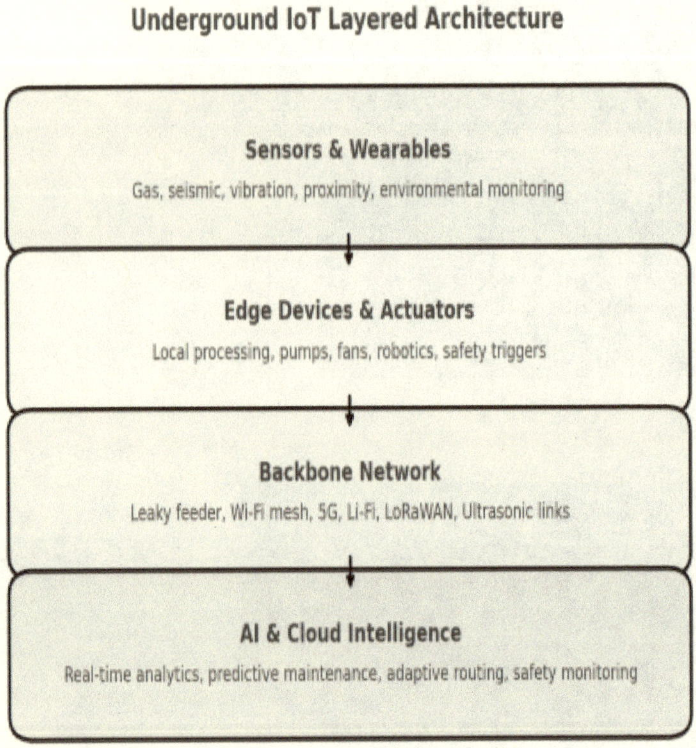

Figure 2: Underground IoT Layered Architecture

This is a diagram of the Underground IoT Layered Architecture. It illustrates the data flow graphically:

Sensors & Wearables → Edge Devices & Actuators → Backbone Network → AI & Cloud Intelligence

3.3. Important Elements of Underground IoT Systems

Underground IoT networks are meticulously planned systems made to endure harsh environments while maintaining constant safety, control, and monitoring. They are not just a collection of unrelated gadgets. Four fundamental pillars—sensors, actuators, communication nodes, and data hubs/gateways—are necessary for these networks to function effectively. They serve as the foundation for mines that are intelligent and adaptable.

3.3.1. **Sensors: The Mine's Eyes and Ears**

Sensors convert physical events into digital insights by capturing the subterranean environment's secret signals. They offer the vital "situational awareness" needed to foresee risks and maximize efficiency.

- **Environmental Measurements**
    - Carbon monoxide, oxygen, methane, and other vital gases are continuously monitored by gas sensors.
    - Sensors for temperature and humidity guarantee worker comfort and identify circumstances that could hasten equipment wear.
    - Micro-shocks that precede rock bursts or collapses are detected by seismic and ground movement sensors.
- **Equipment Performance**
    - Early indicators of imbalance in conveyor systems, fans, or motors are detected by vibration sensors.
    - Hydraulic pressure monitors aid in averting disastrous loader and drill failures.

  - Fuel and engine health sensors optimize consumption, lowering expenses and prolonging equipment life.
- **Location and Health of Personnel**
  - Wearable IoT devices keep an eye on respiration, heart rate, and fatigue levels to make sure employees stay within safe bounds.
  - Real-time visibility of miners is made possible by location tracking systems (RFID tags or UWB beacons), which facilitate prompt emergency response.
  - Hazard exposure sensors track repeated exposure to harmful gasses, dust, or heat.
  - **Example:** BHP's wearable technology Miners in certain BHP operations wear connected gadgets that use micro eye-movement analysis to track fatigue. Before fatigue jeopardizes safety, alerts are set off. Subsequently, sophisticated artificial intelligence algorithms combine streams of data from multiple sensors, identifying irregularities and anticipating malfunctions before they materialize, while sending out automated notifications or activating actuators.

3.3.2. **Actuators: Transforming Information into Action**

Actuators carry out actions, whereas sensors observe. They are the backbone of the subterranean IoT system, carrying out automated reactions to guarantee effectiveness and safety.

- **Ventilation Control**:
  - When gas concentrations increase, fans automatically adjust, bringing fresh air to areas that

need it most and reducing energy consumption through ventilation-on-demand.

- **Managing Materials**
    - Conveyor belts modify their speed to avoid overloading or to align with processes that are upstream or downstream.
    - Hoists and crushers adjust throughput in response to equipment health and real-time demand.
- **Emergency Response**:
    - During dangerous situations, doors and blast barriers automatically close.
    - Robotic evacuation guidance, strobe lights, or alarms are activated to direct employees to safety.
    - A closed-loop system with almost instantaneous detection and response is created by the interaction of sensors and actuators, which lessens the need for human involvement in hazardous situations.

3.3.3. Communication Nodes: The Connectivity Nervous System

Despite the famously challenging underground connectivity, communication nodes serve as the relays and routers that keep the digital mine operational. They guarantee that all signals, including vehicle telemetry and gas data, go to their intended location.

- **Leaky Feeder Systems**
    - Conventional coaxial wires that travel through tunnels to send and receive radio signals. Still useful for dependable voice communication in both directions.
- **Wireless Mesh Networks**

    - A dynamic, self-healing grid of nodes connects them. The network ensures resilience by automatically rerouting traffic in the event of a node failure.
- **Private LTE and 5G solutions**
    - Make ultra-low latency connections possible, which are essential for remote-controlled drilling rigs, drones, and autonomous haul trucks.
    - Offer scalability to manage thousands of concurrently broadcasting IoT devices.
- **Nodes for Edge Computing**
    - Reduce latency and make sure that vital warnings are not dependent just on surface/cloud connectivity by processing data closer to the source.
    - To save bandwidth, filter redundant data and send only insightful information to central systems.
    - **Case Study**: Boliden & Ericsson Autonomous cars and AR headsets are powered by private LTE/5G networks in Sweden's Kankberg mine. By processing data below, edge nodes allow operations to continue even in the event of a brief outage of surface connectivity.

3.3.4. Data Hubs and Gateways: The Network's Brainstem

The gateway layer, which is in charge of gathering, protecting, and sending massive data streams, is at the core of the subterranean Internet of Things.

- **Aggregation:** By combining unprocessed data from hundreds of sensors, gateways eliminate noise and organize data for analysis.
- **Secure Transmission:** Encrypted, fault-tolerant routes must

carry data from the mine's depths to control rooms or cloud platforms on the surface.

- **Connecting to AI Systems:** Gateways connect digital intelligence to the physical mine, providing real-time data to AI-driven analytics for process optimization, hazard detection, and predictive maintenance.
- **Redundancy:** Even in the event of localized failures, dependability is guaranteed by multiple communication channels (wired, wireless, and satellite backup).
- **Example: Vale's Big Data Hubs:** Data is fed into AI-powered digital twins of the mine at Vale's copper operations via gateways. These twins model changing subterranean conditions and suggest flexible drilling and extraction techniques.

3.3.5. **Main Takeaway:** The integrated backbone of the subterranean Internet of Things is made up of sensors, actuators, communication nodes, and gateways. In the most difficult settings on earth, sensors sense, actuators respond, nodes connect, and gateways think, forming a living, adaptable network that improves sustainability, productivity, and safety.

Diagram: AI + IoT = Intelligent Mine

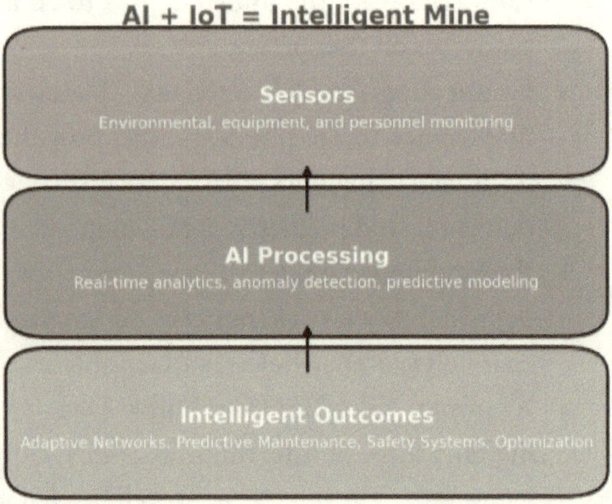

Figure 3: AI + IoT = Intelligent Mine

This is the schematic for the closed-loop underground IoT system. It demonstrates how:

Sensors → Communication Nodes → Data Hubs & Gateways → Actuators → back to Sensors/Environment collaborate in an ongoing cycle, allowing for automated action, real-time monitoring, and decision-making below ground.

Diagram: AI + IoT = Intelligent Mine (2)

Figure 4: AI + IoT = Intelligent Mine (2)

The final layer has been split into **four distinct intelligent outcomes**:

- **Adaptive Networks**
- **Predictive Maintenance**
- **Safety Systems**
- **Optimization**

3.4. AI and IoT: The Power of Intelligent Connectivity

Artificial Intelligence (AI) offers the brain that interprets this deluge of data, while the Internet of Things (IoT) provides the mine's eyes, ears, and touch points. IoT devices alone produce useful streams of raw data, but a large portion of this data is underutilized in the absence of intelligence. *AI turns that data into insights that may be put to use, resulting in a self-learning, self-optimizing system that changes as the mine does.* The intelligent mine, which not only reacts to events but also foresees and stops them, is built on the foundation of AI and IoT.

3.4.1. **Adaptive Networks: Intelligent Connectivity**

- **The Problem:** Because of signal absorption, interference, and moving tunnels, underground networks are inherently unstable. These erratic shifts are too much for conventional communication infrastructure to handle.
- **The Advantage of AI:**
    - Algorithms driven by AI continuously track network traffic loads.
    - Potential congestion is identified before it occurs using predictive models.
    - Dynamically generated routing decisions guarantee continuous connectivity even in areas with disruptions.
- **Outcome:**
    - A self-healing network with reduced downtime and constant delivery of vital information to decision-makers, such as worker location signals and gas alerts.
- **Example:** During **Ericsson's testing with mining partners**, AI-powered 5G systems instantly redistributed bandwidth to

self-driving trucks, guaranteeing that no data was lost even during periods of increased demand.

3.4.2. Transitioning from Reactive to Proactive Predictive Maintenance

- **The Challenge:** Subterranean equipment failures result in expensive delays, safety hazards, and can necessitate challenging, time-consuming repairs in tight areas.
- **The Advantage of AI:**
  - Equipment vibration, temperature, hydraulic pressure, and acoustic characteristics are recorded by IoT sensors.
  - AI models identify patterns such as overheating, misalignment, and bearing wear that occur before breakdowns.

    By scheduling maintenance prior to a catastrophic failure, downtime can be minimized and asset life can be increased.
- **Outcome:** Switching from reactive firefighting to predictive precision maintenance guarantees uninterrupted operation and saves millions of dollars yearly.
- **Case Study:** Rio Tinto Rio Tinto significantly increased fleet availability by reducing unscheduled maintenance events by almost 30% through the use of AI-enhanced predictive analytics on haul trucks.

3.4.3. Intelligent Safety using Dynamic Safety Systems

- **The Challenge**: Gas explosions, rock falls, fires, and flooding are among the inherent hazards of underground mining. Conventional safety measures are frequently slow and reactive.

- **The Advantage of AI:**
    - Continuous data on water levels, earthquake activity, and air quality is fed via environmental IoT sensors.
    - Anomalies, such as methane spikes or strange vibrations, are promptly interpreted by AI models.
    - Automatic safety procedures are initiated, such as activating barriers, turning on fans to improve ventilation, setting off alarms, or lighting evacuation pathways.
- **Outcome:** A mine that can foresee risks and take preventative action in real time, frequently more quickly than human managers could.
- **Example:** South African gold mines have implemented AI-driven gas monitoring systems that, upon detecting hazardous quantities, instantly cut off electricity to high-risk areas.

3.4.4. Operational Optimization: Using Intelligence to Increase Efficiency

- **The Challenge**: Mines are enormous, intricate systems with thousands of moving components, including electrical networks, ventilation systems, machinery, and humans. Inefficiency is unavoidable in the absence of intellect.
- **The Advantage of AI:**
    - To cut down on idle time, real-time analytics schedule trucks, loaders, and drills.
    - AI supplies air only where and when needed, balancing energy efficiency with ventilation demands.
    - By coordinating industrial processes, predictive algorithms guarantee that ore extraction, transporting, and processing go without a hitch.
- **Outcome**: Leaner, safer, and more energy-efficient operations

that produce more with a smaller carbon footprint.

- **Case Study: Boliden's Kankberg Mine**: IoT-powered AI-assisted ventilation-on-demand improved worker comfort and safety while reducing energy use by 40%.

3.4.5. **From Responsive to Anticipatory Mines**

The biggest benefit of integrating AI and IoT is not just solving issues but also foreseeing them.

- My approach to crisis management has changed *from reactive to proactive.*
- Systems develop into *cognitive ecosystems* that can learn from past data and adjust to novel geological or operational circumstances.
- The end result is *a living, breathing mine that keeps getting better every year, boosting sustainability, production, and safety.*

3.4.6. **Important Takeaway:** The mine is no longer a static, dangerous place when AI and IoT meet underground. Rather, it transforms into an ecosystem that is intelligent, adaptive, and predictive—a mine that sees, thinks, and responds instantly. This collaboration aims to redefine mining's fundamental characteristics rather than merely focusing on technology.

3.4.7. **Case Study: Australia's Autonomous Underground Mine**

More than 2,000 Internet of Things sensors were placed across the subterranean network of a state-of-the-art gold mine in Australia:

- Low-latency, high-speed communication was made possible by wireless mesh networks.
- Analytics driven by AI were 92% accurate in predicting machinery problems.
- A 38% decrease in danger incidences was achieved by automatic ventilation adjustments and emergency notifications.
- **Outcome:** The end result is an intelligent, fully networked underground operation that can coordinate people, machines, and the environment in real time.

3.5. The Path Ahead: Extremely Interconnected Mines

The history of mining has always been one of resource extraction—going farther, digging deeper, and enduring more difficult circumstances. The story of the future, however, is one of autonomy, intelligence, and connectedness. In addition to machinery and labour, the mining industry of the future will depend on hyper-connected ecosystems in which AI, IoT, and next-generation communication networks seamlessly integrate every sensor, machine, worker, and decision-making system.

The industry's digital transformation, which has made underground surroundings that were previously thought of as "black boxes" transparent, intelligent, and adaptable, is exemplified by the hyper-connected mine.

3.5.1. Seamless Communication Infrastructure

- **Next-Gen Networks:** Deep underground high-speed, low-latency coverage will be extended by 5G, mesh Wi-Fi, and private LTE systems.
- **Self-Healing Connectivity**: Signals are automatically rerouted around obstructions, equipment interference, or power outages thanks to AI-driven network orchestration.
- **Unified Data Flows**: All data, including geospatial maps, equipment health data, and worker biometrics, are continuously incorporated into a unified digital ecosystem, eliminating the need for silos.
- **Vision:** A mine where connectivity is always available, even kilometres below the surface, allowing autonomous systems and people to coordinate in real time.

3.5.2. Ecosystems for Cognitive Mining

- **AI as thc "Brain":** IoT sensors gather millions of data points every second, ranging from micro-seismic impulses in rock to vibration in drills. This is combined with AI to create real-time situational awareness.
- **Digital Twins:** Each mine has an in-the-moment digital mirror that models production procedures, safety situations, and geological shifts before they happen.
- **Self-Optimizing Systems:** fleets reorganize themselves to avoid congestion, ventilation changes dynamically, and machines automatically schedule based on ore flow.
- **Result:** Mines that are able to reason, anticipate, and change in real time in addition to simply "operating."

3.5.3. The Partnership Between Humans and Machines

- **Connected Workers**: Wearable technology improves safety and lowers accident rates by tracking worker location, health, and environmental exposure.
- **AI Assistants Underground**: Augmented reality (AR) helmets provide workers with real-time supervision by presenting escape routes, ideal drilling angles, and hazard alarms.
- **Collaborative Autonomy:** AI handles everyday tasks while humans oversee fleets of self-driving trucks, loaders, and drones, intervening only in rare circumstances.
- **Impact:** With the help of machine intelligence, humans are now strategic controllers and decision-makers rather than being replaced.

3.5.4. Sustainability via Interaction

- **Energy Optimization**: AI reduces wasted energy by dynamically balancing power between processing, hauling, and ventilation.
- **Carbon-Aware Mining**: Real-time emissions measurements from connected mines enable prompt modifications to reduce environmental effect.
- **Circular Mining**: By ensuring that by-products are collected, recycled, or reused, IoT-enabled waste tracking helps to lessen the environmental impact.
- **Vision:** Mines that support global sustainability objectives and conserve ecosystems in addition to extracting resources.

3.5.5. Resilience and Security

- **Cybersecurity Backbone**: Threats grow along with connection. Critical mining infrastructure will be protected from cyberattacks by AI-driven anomaly detection.
- **Resilient Design:** Distributed, redundant systems will be used in hyper-connected mines to ensure that activities continue without interruption even in the event of a node failure.
- **Disaster Preparedness**: Mines can use predictive analytics to anticipate hazards like power outages, seismic shifts, and flooding, which sets off automated backup measures.

3.5.6. Moving Toward Completely Self-Reliant Mines

The autonomous mine is the ultimate horizon:

- **No-Human Zones:** To lessen exposure to risks, entire mining sections could function without any subterranean workers.
- **Autonomous Fleets**: Conveyors, drills, loaders, and trucks that work together without human assistance.
- **Remote Command Centres**: Using immersive digital twins,

human supervisors hundreds of kilometres away supervise activities.

- **Case Study: Sandvik and Rio Tinto:** An early example of hyper-connected autonomy can be seen in the Pilbara region, where autonomous mining trucks are already controlled by remote operation centres in Perth thousands of kilometres away.

3.5.7. Important Takeaway:

The path forward leads to *intelligent, hyper-connected mines*—living, breathing digital ecosystems where communication, IoT, and AI technologies meet. Compared to the industrial past, these mines will be safer, more environmentally friendly, more productive, and more robust. Extracting intelligence from data will replace the previous focus on extracting value from the ground.

3.6. A Day in the Life of a Hyper-Connected Mine (Year 2035)

Even though no one has gone underground yet, the mine is already awake at 06:00hrs in the morning of 2035.

Operators access immersive digital twin interfaces ***from a distant command centre hundreds of kilometres away.*** *Every shaft, tunnel, and machine is visible in real time on the mine's holographic projection. A ventilation anomaly in the eastern tunnel and an anticipated rock stress zone close to Shaft 17 are highlighted by AI helpers as areas of attention.*

6:30 a.m.: Change of Shift

Unlike their ancestors, miners do not instantly descend into the darkness after arriving on the scene. Rather, they wear ***AR helmets and wearable health monitors.*** *AI monitors their vitals, including heart rate, tiredness, and hydration, to make sure no one starts a shift in a dangerous state. After then, positions are dynamically given to workers according to their experience, health, and current mine requirements.*

7:00 a.m.: Machines Take the Lead

In Zone 4, autonomous drilling rigs start up. Feedback on drill angles, pressure, and vibration is provided by Internet of Things sensors embedded in the rock. To avoid stress fractures, the AI system modifies the drilling parameters in milliseconds. Meanwhile, a network of subterranean 5G nodes continuously optimizes the trajectories of a convoy of ***autonomous haul trucks*** *that transport ore to processing facilities in perfect synchronicity.*

9:15 a.m.: Prioritize Safety

Near the western tunnel, a sensor network detects an increase in methane levels. ***AI safety measures are activated instantly****:*

- *Fans for ventilation reroute airflow.*
- *The impacted area is sealed off by automated doors.*
- *Workers are guided to safety by evacuation arrows displayed on their AR helmets.*
- *A real-time update is sent to the command centre above ground. The danger is eliminated in a matter of minutes, resulting in no injuries, bewilderment, or fear.*

11:00 a.m.: Predictive Maintenance

An IoT-enabled loader in the equipment bay warns that its hydraulic pump will break down in 36 hours. AI reassigns work to other loaders, orders the new part through the supply chain system, and plans downtime for the next time when demand is low. With just one click, human supervisors give their approval, avoiding expensive delays and unplanned malfunctions.

1:30 p.m.: AI + Human Cooperation

To sample the geology, one crew descends. They use AR overlays that display ***real-time seismic data, ore distribution estimates, and recommended drilling locations in place of human mapping****. Although AI produces suggestions, human judgment ultimately decides the matter. AI gives accuracy and foresight, while people contribute creativity and intuition, resulting in a seamless collaboration.*

3:45 p.m.: Sustainability in Action

AI-driven grid balancing and solar-powered energy storage maximize electricity between processing, transportation, and ventilation. Emissions monitors provide communities and authorities with real-time data that demonstrates adherence to carbon targets. Waste is automatically scanned, segregated, and recycled. The mine is ***environmentally conscious*** *in addition to being productive.*

6:00 p.m.: Shift End, Mine Continues Running

The mining goes on as employee's clock out. Ore is processed, drilled, and hauled by autonomous fleets at night. Every parameter is monitored by AI, and subsurface edge computer nodes maintain operations even in the event of a momentary outage in surface connectivity.

A supervisor in the command centre goes over the day's report, which includes ***zero accidents, maximum output, lower energy use, and forecasted information for operations tomorrow.***

The hyper-connected mine of the future*, where data, intelligence, and human skill combine to form a single, ever-evolving system, is not science fiction.*

Chapter 4: AI-Driven Signal Intelligence: Enhancing and Guaranteeing Data Flow

4.1. Smooth Underground Communication

AI-powered signal intelligence acts as the brain and central command if IoT networks are the mine's nervous system. It *continuously analyses, forecasts, and adjusts to guarantee that communication never breaks.* Underground mines are surrounded by both natural and man-made obstacles, such as winding tunnels, uneven rock densities, vibrations from heavy machinery, electromagnetic interference, and excessive humidity or dust, in contrast to open-surface settings where radio waves can pass easily. Conventional communication methods are insufficient in this maze. AI turns brittle subterranean connections into strong, intelligent lifelines by introducing the capacity to learn, anticipate, and self-correct.

4.1.1. Intelligent Routing of Signals

AI-powered algorithms dynamically choose the optimal routing for data packets based on ongoing network state monitoring. *When a node fails, the system self-optimizes like a living thing, rerouting traffic around obstructions and preserving uptime in place of inflexible routes.* This is especially important when it comes to telemetry data from autonomous trucks and drilling rigs or emergency voice communication.

- **Example**: AI-managed mesh networks showed 40% fewer communication dropouts than static routing systems in a **Canadian Nickel Mine,** guaranteeing continuous underground vehicle coordination.

4.1.2. Management of Predictive Interference

There are many competing signals in underground environments, ranging from autonomous machines to blasting equipment. AI algorithms can anticipate when interference is likely to occur and make proactive adjustments to frequencies or transmission power levels based on prior interference patterns.

- **Example**: An AI system can identify when scheduled blasting in a certain shaft will interfere with Wi-Fi signals and redirect data traffic through leaky feeder systems to avoid downtime.

4.1.3. Flexible Allocation of Bandwidth

Not all buried data is created equal. Routine system logs must never take precedence over real-time safety alerts, gas detection readings, or worker tracking data. AI allocates bandwidth where it is most needed by giving traffic priority levels in real time. This guarantees that, even during traffic jams, safety-critical information is never delayed.

4.1.4. Signal Improvement and Regeneration

Weak signals can be rebuilt with little loss thanks to AI-enhanced error correction and signal regeneration techniques. Clarity in voice and sensor transmissions is ensured by the system's ability to "fill in" lost data packets with high accuracy thanks to machine learning algorithms trained on patterns of signal deterioration in various rock formations.

4.1.5. Communication Aware of Context

Signal intelligence driven by AI does more than just handle data mindlessly; it also understands context. For example, the system automatically prioritizes hazard alerts across all channels, temporarily reducing non-essential traffic, if gas sensors identify rising methane levels. Telemetry streams from impacted machinery are also given priority for diagnosis during equipment failures.

4.1.6. Networks for Self-Healing and Fault Prediction

AI foresees mistakes before they occur. The system finds weak points in the communication network by keeping an eye on trends in packet loss, latency spikes, or node performance. In order to avoid downtime, it can then proactively reroute data, reassign channel frequencies, or even suggest physical action, such as moving antennas.

4.1.7. AI-Human Collaboration in Signal Control

Humans maintain oversight while AI manages the complexity. AI analytics-powered visual dashboards that provide network health, risk forecasts, and improvement suggestions are provided to engineers. Operators switch from addressing communication breakdowns to strategic decision-making, depending on AI's predictive ability to stay ahead of interruptions.

4.1.8. The Need for Safety

In underground crises, every second matters. During emergencies, such as roof collapses, equipment fires, or poisonous gas leaks, AI-enabled signal intelligence makes sure that emergency messages are prioritized immediately, evacuation orders are sent out right away, and rescue teams stay in real-time communication even in areas that have been compromised.

4.1.9. Towards Cognitive Mines

In the future, AI signal intelligence will develop into cognitive communication ecosystems rather than just reactive optimization. These systems will:

- *Gain knowledge from each event to enhance performance in the future.*
- *Easily integrate with digital twin mines to mimic communication breakdowns before they occur.*
- *Make conversational interfaces between humans and AI possible.*

This will allow miners to communicate directly with the network by asking questions like *"Where is the nearest safe exit?"* or *"What is the current gas level in Shaft 7?"* and getting prompt, AI-driven answers.

4.1.10. **Key Takeaway:**

AI-powered signal intelligence guarantees that underground mining communication is not only dependable but also robust, adaptive, and lifesaving, much like the human brain filters noise, prioritizes attention, and adjusts to constantly changing circumstances. Deep beneath the surface of the planet, it turns communication systems from passive infrastructures into proactive defenders of efficiency and safety.

4.2. The Challenge: Latency and Signal Loss

One of the most challenging settings for communication networks is an underground mine. Conventional technologies, such as cable connections, leaky feeder networks, and fixed radio frequencies, can perform well in stable environments, but they are severely constrained by the dynamics and physics of underground operations. Safety, effectiveness, and operational decision-making are all directly impacted by these constraints, which in reality show up as signal loss, delay, and erratic network failures.

4.2.1. Attenuation of Signals

Signals traveling through tunnels are weakened by the natural absorption, scattering, or reflection of radio waves by rocks, soil, and metal infrastructure. A new excavation or a slight change in tunnel direction can result in "dead zones" where communication is impossible. Because traditional systems are not intelligent enough to adapt to these changes dynamically, personnel and equipment are disconnected at crucial times.

4.2.2. Congestion in the Network

For real-time monitoring, modern mines are depending more and more on wearable technology, IoT sensors, networked machinery, and driverless cars. Critical telemetry data may be delayed, packets may be missed, and latency may increase when hundreds or even thousands of devices vie for the same limited bandwidth. Because static setups are unable to dynamically reorder traffic, operational directives may be carried out and safety alarms may arrive too late.

4.2.3. Interference from the Environment

The subterranean environment is hostile and uncertain. *Signal quality is deteriorated by vibration from heavy gear, dust clouds from drilling, extreme humidity, and even electromagnetic interference from mining equipment.* Conventional systems are mostly passive and unable to adjust to changing circumstances, which leads to sporadic connectivity and inconsistent operation.

4.2.4. Adaptive Topology

Mines are dynamic, living things. Temporary support structures are built, additional shafts are dug, and tunnels are expanded. Static networks are unable to adjust to these changes on their own. A network node that is positioned ideally one month can stop working the next, resulting in blind spots that jeopardize data flow, machine coordination, and safety.

Diagram: The Challenge: Signal Loss and Latency

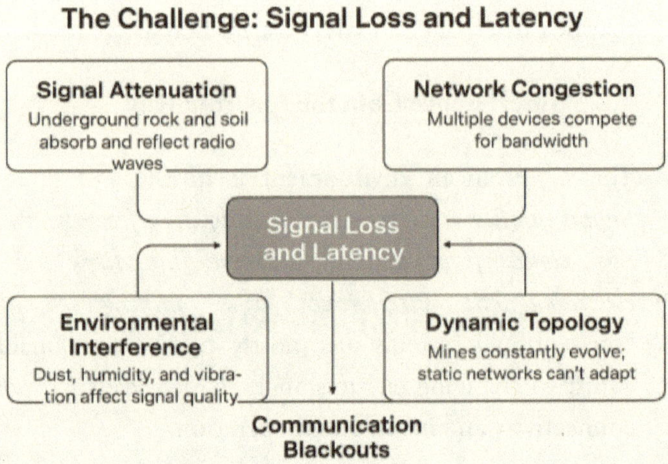

Figure 5: The Challenge: Signal Loss and Latency

4.3. Implications of Unmanaged Signal Issues

These technological constraints result in real operational hazards in the absence of AI-driven intervention:

- **Communication Blackouts**: Unexpectedly, workers, cars, and sensors lose connectivity.
- **Delayed Alerts**: It's possible that staff crises, equipment malfunctions, or gas leaks won't be communicated right away.
- **Operational Inefficiencies**: IoT-based monitoring systems may offer partial or erroneous insights, and autonomous cars may halt or collision.
- **Safety Risks**: In dire situations, poor communication can result in harm or death, particularly when emergency cooperation is crucial.

Traditional subterranean communication systems are essentially static, reactive, and brittle by nature. They find it difficult to stay up with the fast-paced, dangerous subterranean environment; AI-powered signal intelligence is well positioned to bridge this gap.

AI turns subterranean networks from brittle infrastructures into robust, intelligent, and proactive communication ecosystems by continuously monitoring, forecasting, and adjusting to changing situations.

4.4. Intelligent Signal Management: AI's Rescue

One of the most difficult settings for communication networks is an underground mine. Signal attenuation, changeable layouts, and physical obstacles are problems for traditional systems. However, *AI changes this environment by bringing in autonomous optimization, predictive intelligence, and adaptability*. AI's ability to redefine underground communication is demonstrated by the following capabilities:

4.4.1. Allocation of Dynamic Frequencies

Fixed frequency channels, which are frequently used by conventional radio systems, are susceptible to interference from objects such as rock formations and machinery. This paradigm is altered by frequency management driven by AI:

- **AI-Powered Decisions:** AI continuously tracks network utilization, interference trends, and signal quality. In order to maintain optimal connectivity, it dynamically reallocates frequencies and anticipates congestion before it occurs.
- **Real-World Example:** By automatically changing channels in response to identified interference from high-voltage equipment, AI-controlled radios in a contemporary underground copper mine decreased signal dropouts by more than 40%.
- **Prospects for the Future:** As AI systems advance, they may be able to incorporate spectrum sharing among several mines, enabling regional mining networks to efficiently coordinate frequencies.

4.4.2. **Predictive Routing**

Because mining operations cause underground tunnels to change often, fixed routing frequently results in delays or data loss. Predictive routing powered by AI solves this issue via:

- **Predicting Network Congestion:** AI algorithms use real-time sensor data and historical traffic patterns to forecast potential areas of congestion or signal obstruction.
- **Adaptive Data Flow:** AI redirects packets via different pathways when it detects a possible bottleneck, guaranteeing that voice communications, video feeds, and vital telemetry continue unhindered.
- **Case Study:** AI predictive routing allowed autonomous trucks in a South African gold mine to keep up constant telemetry even during extensive blasting operations, avoiding expensive downtime.
- **Future Vision:** Self-optimizing networks that automatically distribute load among several communication modalities, such as Wi-Fi, 5G, and LoRaWAN, may develop from predictive routing.

4.4.3. Networks That Heal Theirs

Underground communication can be severely hampered by node failures or environmental changes, such as collapsed tunnels, dust clouds, or the temporary installation of machinery. AI's capacity for *self-healing* builds resilience:

- **Automated Recovery**: Without human assistance, the network automatically reroutes traffic through backup nodes in the event that a communication node fails, preserving integrity.

- **Intelligent Redundancy**: AI anticipates communication problems by identifying key nodes whose failure would most affect it and proactively creates fallback routes.
- **Example:** self-healing mesh networks were installed in an Australian coal mine. In the event of an unscheduled collapse, the network autonomously rearranged itself to maintain complete safety systems.
- **Looking Ahead:** AI networks of the future may be able to anticipate possible node failures before they happen, triggering preventive actions like drone-based relay units or temporary mobile nodes.

4.4.4. **Reduction of Latency**

Every millisecond matters in subterranean settings, particularly for robotic equipment, autonomous cars, and real-time hazard alarms. *Latency is significantly decreased when edge computing and AI are used together*:

- **Edge Processing**: Instead of being transmitted to surface servers, critical calculations, like gas detection warnings or vehicle path planning, are handled locally at the network edge.
- **Immediate Response**: AI guarantees automated machinery receives feedback instantly, reducing the risk of crashes, streamlining routes, and enhancing general safety.
- **Example:** edge AI allowed autonomous loaders at a platinum mine to react to unexpected impediments in milliseconds, increasing productivity and security.
- **Prospects for the Future**: Edge AI may develop into decentralized "mini brains" for every network node, enabling real-time machine and human collaboration across the mine,

even in places with no surface connectivity.

In conclusion, artificial intelligence (AI) turns subterranean communication from a brittle, static system into a robust, intelligent network that can foresee problems, fix itself, and respond instantly. As machine learning models advance, completely autonomous, adaptive mining networks are anticipated in the future, with AI handling communication, safety, and operational efficiency while human operators concentrate on making decisions.

4.5. Integration of AI and IoT: The Underground Power of Intelligent Connectivity

The full potential of underground communication networks is unlocked by the combination of artificial intelligence (AI) and the Internet of Things (IoT). *IoT guarantees a steady flow of data from each device, sensor, and vehicle in the mine, while AI controls signal integrity.* When combined, they produce an intelligent, self-optimizing, and predictive hyperconnected ecology.

4.5.1. **Environmental Awareness in Real Time**

- **IoT Sensors across the Board**: Data is continuously streamed by vibration monitors, temperature sensors, humidity sensors, and gas detectors.
- **AI-Driven Insights**: AI examines this inflow in real time, identifying irregularities, forecasting dangerous situations, and sending out warnings before threats become more serious.
- **Example:** AI combined with Internet of Things gas sensors in a platinum mine identified hazardous methane build-up minutes before human notice, enabling automated ventilation modifications.

4.5.2. **Predictive Upkeep**

- **Continuous Equipment Monitoring**: Wear levels, vibration, and motor temperature are all monitored via IoT sensors.
- **AI Forecasting:** By using machine learning to anticipate malfunctions before they occur, maintenance is scheduled proactively as opposed to reactively.
- **Outcome:** Lower operating expenses, longer equipment

lifespan, and less downtime.

- **Case Study:** After using AI-IoT predictive maintenance solutions, a copper mine in Chile reported a 30% decrease in unscheduled machinery downtime.

4.5.3. **Autonomous Fleet Coordination**

- **Internet of Things Vehicle Tracking**: Telemetry and positioning sensors are installed on trucks, loaders, and drills.
- **AI Traffic Control**: AI-managed signal allocation and predictive routing provide smooth operation of autonomous cars, preventing bottlenecks and collisions.
- **Example:** AI managed a fleet of self-driving haul trucks in a South African gold mine, maximizing haulage cycles and increasing fuel efficiency by 15%.

4.5.4. **Systems of Dynamic Safety**

- **Instant Hazard Response:** AI uses IoT sensor data, such as temperature anomalies, gas spikes, or structural changes, to initiate automated safety procedures including ventilation adjustments, alarms, and evacuations.
- **Learning from Data:** Every event enhances AI models, increasing the precision of hazard prediction and reaction in the future.
- **Looking Ahead:** Wearable sensors on miners may be integrated into the next generation of AI-IoT systems, allowing for customized safety alerts depending on factors like location, exposure levels, and fatigue patterns.

4.5.5. **Optimization of Operations**

- **End-to-end Automation:** To optimize efficiency, AI

coordinates environmental controls, vehicle movement, network traffic, and maintenance plans.

- **Real-Time Decision Support** allows human operators to concentrate on strategic choices rather than routine monitoring by providing them with actionable insights rather than raw data.
- **Future Vision:** Fully automated mines with AI-IoT networks that continuously learn from data, improve resource allocation, and oversee daily operations while guaranteeing complete safety.

In conclusion, subterranean mines become intelligent ecosystems as a result of the confluence of AI and IoT. Each sensor and gadget adds to the continuously advancing operational intelligence, and communication networks become durable, adaptive, and predictive. In this setting, productivity, efficiency, and safety are engineered, optimized, and automated rather than reactive.

4.5.6. **Case Study: Platinum Mine in South Africa**

AI-driven signal intelligence was applied throughout the underground operations of a platinum mine in South Africa:

- An 87% reduction in signal dropouts was achieved by dynamic frequency allocation.
- All staff received emergency alerts in a matter of seconds, enhancing safety reaction times; predictive routing made sure that autonomous loaders received real-time orders without delays.

The mine turned communication from a weakness into a strategic advantage by fusing AI, IoT, and digital twin technologies.

4.5.7. **Cognitive Communication Networks in the Future**

Subterranean networks driven by AI will develop into fully cognitive systems in the future:

- **Self-learning networks** that adjust continuously in response to changes in the environment and usage patterns.
- **Automatic network reconfiguration** that eliminates downtime for the installation of new tunnels or equipment.
- **Predictive hazard communication**, in which warnings are sent out prior to the occurrence of hazardous situations.
- **Cloud and 5G/6G network integration**, which allows for smooth communication between remote operations centres and subterranean areas.

Signal intelligence driven by AI is not merely a fix; it is the cornerstone of the future's autonomous, highly networked mines. It guarantees accurate, safe, and effective communication, coordination, and operation of the mine regardless of its complexity.

Chapter 5: Self-Organization and Communication: How Humans and Machines Collaborate Effortlessly Underground

Overview

For many years, underground mining was characterized by isolated activities and disjointed communication. Human miners operated in isolated areas with little insight into one another's job, and machines frequently operated on their own with only manual instructions. Inefficiency, redundant work, and preventable dangers were the outcome.

Subterranean operations are now being transformed into cooperative ecosystems where humans and machines synchronize flawlessly thanks to AI-driven autonomous systems and IoT-enabled communication networks.

In the future underground mine, neither independent machines nor solitary humans are employed. Rather, a symbiotic ecology develops in which personnel and autonomous machinery may work together seamlessly thanks to AI and IoT. In order to maximize efficiency, productivity, and safety, this chapter examines how intelligent communication networks coordinate this intricate dance between humans and machines.

5.1. The Role of Autonomous Coordination

The goal of autonomous coordination is to establish shared situational awareness among humans, robots, and AI-driven systems, not just to have machines operate without human input. AI serves as the "orchestrator," directing decision-making and workflows across several mining activity tiers:

- **Task Distribution**: AI uses real-time data to allocate tasks to robotic arms, drills, and autonomous trucks.
- **Workflow Synchronization**: Mechanisms synchronize motion to prevent machines from interfering with one another or posing a risk to human operators.
- **Adaptive Responses**: AI rapidly adjusts schedules and communication channels in response to changing circumstances, such as unanticipated rock falls or equipment failures.

5.1.1. Case Study: Real-Time Haulage Coordination

A single AI hub facilitates communication between a fleet of self-driving haul trucks in a sizable underground copper mine. The AI immediately reroutes neighbouring cars to avoid delays and accidents when a rock fall momentarily blocks a transport road. From the surface control room, human supervisors keep an eye on the changes, but they are not required to step in because the system resolves the disturbance on its own.

5.1.2. Human-Machine Cooperation

Although they still play a crucial role underneath, humans' responsibilities are changing from manual labour to strategic and managerial duties:

- **Augmented Decision-Making**: AI helps engineers make better decisions by providing safety analytics and quality projections.
- **Safety Guardianship**: Wearable technology monitors a miner's heart rate, location, and exposure to harmful gasses. AI notifies users and initiates ventilation adjustments when thresholds are exceeded.
- **Collaborative Interfaces**: While working below, miners can view digital overlays like geological models or equipment diagnostics thanks to AR headsets.

5.1.3. The Augmented Safety Officer Scenario

The oxygen levels of a miner wearing an AI-linked helmet abruptly plummet. This is immediately communicated to the network by the helmet. In response, the AI does the following:

1. *Notifies autonomous drones in the area to look for the miner.*
2. *Increasing airflow by rerouting ventilation.*
3. *Using an AR overlay to send directions to the miner so they can leave safely.*

All of this takes place in less than ten seconds, which is faster than conventional manual communication chains.

5.2. Machine -to - Machine Communication (M2M)

In underground cooperation, machine-to-machine communication is among the most innovative features:

- **Vehicle Convoys**: To avoid collisions in small tunnels, autonomous loaders and trucks exchange location data.
- **Dynamic Routing:** Drills use dynamic routing to modify their patterns in order to prevent vibration interference with adjacent equipment.
- **Resilient Networking**: To prevent a communication blackout in the event of a node failure, mesh networks reroute signals.

Case Study: Predictive Drill Optimization

Two self-sufficient drills working in nearby shafts exchange real-time vibration and stress data. One drill will alert the other to change torque and angle if it senses too much pressure. This lowers the chance of a collapse by avoiding undue stress on the rock face.

5.3. Constructing an Digital Nervous System

The digital nervous system of underground mines is made up of AI, IoT, and sophisticated communication protocols:

- **Perception:** Sensors collect biological, mechanical, and environmental information.
- **Processing:** Patterns and anomalies are detected by AI.
- **Action:** Coordinated responses are carried out by machines, which may reroute, modify, or shut down as necessary.

The Rock Burst Response Scenario:

Subterranean sensors detect a sudden micro-seismic event. AI anticipates the possible collapse of a tunnel in milliseconds:

- Operations are stopped by autonomous drills.
- Drones and robots are sent out to evaluate the damage.
- Miners are directed to safety exits by AR alerts.
- The mine's computerized nervous system is responsible for keeping the incident under control without causing any harm or equipment loss.

5.4. The Path Ahead: Toward Completely Self-Sustained Ecosystems

Self-managing ecosystems where human-machine cooperation is smooth are the ideal scenario for underground communication:

- **Zero-Harm Mines:** Risks are eliminated before they materialize thanks to predictive AI.
- **Real-Time Co-Adaptation:** This reduces downtime by having machines dynamically adapt to human demands.
- **Invisible Communication**: Because the system maintains connectivity in the background, people no longer consider it.

5.4.1. Future Situation: The 2035 Symbiotic Mine:

A single surface operator uses a holographic dashboard to control a fleet of 200 autonomous equipment operating underground in a future platinum mine. Robotic swarms manage rock sampling and upkeep, while AI plans drilling, haulage, and ventilation. Underground miners no longer rely on radios or hand signals; instead, all commands, alerts, and updates pass through the network undetected. Machines manage execution; humans concentrate on strategy and supervision.

In this scenario, the deep mine is a living, breathing organism rather than a collection of separate actors, where machines and humans work together in symbiosis, each enhancing the other's advantages.

5.5. The Rise of Autonomous Mining

Today, autonomous cars and robotic systems are revolutionizing mining operations; they are no longer sci-fi ideas. Globally, progressive mining firms are using fleets of machines that require little to no human involvement, significantly increasing productivity, safety, and efficiency.

- **Autonomous Haulers and Loaders**: These days, *loaders and self-driving trucks move ore with extreme precision, cutting down on delays brought on by accidents, shift changes, and human fatigue.* AI-driven navigation algorithms steer these machines, enabling them to adjust to subterranean road conditions, steer clear of crashes, and maximize efficiency by optimizing routes.

As a result, there are fewer operational disruptions and reduced fuel usage during continuous, 24-hour operation.

- **Automated Excavators and Drills**: Once reliant on knowledgeable human operators, *precision excavation and drilling are now carried out by clever, automated machinery.* AI-guided drills maximize mineral recovery while minimizing waste by optimizing drilling patterns based on geological data. In order to account for changes in rock density and structural stability, excavators that are outfitted with machine vision and sophisticated sensors modify their digging methods in real time. By minimizing wear and tear, this accuracy not only boosts output but also prolongs the life of mining machinery.
- **Drones and Inspection Robots**: Ground-based inspection robots and unmanned aerial drones are revolutionizing mine monitoring. Long before they endanger workers, *they map*

tunnels in three dimensions, locate infrastructure weak areas, and uncover environmental concerns like gas leaks or rock instabilities. Operational safety is further improved by predictive maintenance and proactive hazard management made possible by real-time data streams from these devices.

However, only when autonomous mining systems are properly integrated will their full potential be realized. Every operator, sensor, and machine needs to be incorporated into a *strong and smart communication system.* Drills may lose synchronization, autonomous haul trucks may stall, and inspection robots may not send out vital hazard signals if there is unreliable connectivity.

Therefore, the development of underground communication systems is inextricably linked to the emergence of autonomous mining. As the mine's central nervous system, high-bandwidth, low-latency networks facilitate real-time decision-making, coordinate machine-to-machine (M2M) interactions, and guarantee that human supervisors maintain control over a highly automated environment. Communication will become the *lifeblood of the future intelligent mine*, not merely a facilitator, as autonomy increases.

5.6. Autonomous Mining's Future: Going Beyond Current Potential

The future holds an even bigger leap—toward self-organizing, adaptive mining ecosystems driven by artificial intelligence, robots, and next-generation communication networks—even if autonomous loaders, haulers, drills, and drones are already changing operations:

- **Mining Swarm Robotics**: Instead than depending only on large single-purpose machines, picture fleets of smaller, more nimble mining robots operating in synchronized "swarms." These swarms of robots might spread out throughout tunnels, drilling, excavating, moving, and scanning at the same time. In order to maintain operations, other robots dynamically modify their responsibilities in the event that one fails. Highly adaptable, robust, and scalable mining systems that can quickly adjust to changes in the environment or geology are made possible by swarm intelligence, which is based on ant colonies and bee hives.
- **AI-Driven Coordination Platforms:** Advanced AI platforms will coordinate all mining activity in real time, eliminating the need for humans to micromanage autonomous equipment. With little assistance from humans, these "digital mine managers" will distribute work among machines, maximize energy efficiency, and coordinate activities. Long before humans would notice an issue, an AI platform might, for instance, automatically slow down transporters to match excavation speed or reassign drones to monitor parts exhibiting structural instability.
- **Self-Healing Mines:** Inspection robots of the future might be able to actively fix dangers in addition to detecting them. These systems might self-replace damaged parts, reinforce

weakened walls, or fix fractures if they were outfitted with robotic arms, additive manufacturing equipment, or sealants. Mines would transform into self-maintaining structures, drastically increasing worker safety and decreasing downtime.

- **Human-Robot Cooperation**: The human role is redefined by autonomy, not eliminated. Through immersive digital twins, human operators will monitor mining ecosystems, supervising and intervening as needed via *Augmented Reality* (AR) or *Virtual Reality* (VR) interfaces. Workers will direct mining operations from control centres at the surface level, dealing with AI systems like mission commanders instead of being subjected to subsurface hazards.
- **Autonomous Supply Chains**: Autonomy goes beyond the concept of extraction. A completely automated supply chain might incorporate processing facilities, stockpile management, and even ore transportation. Conveyor systems and autonomous trucks would transport ore to processing facilities with ease, where AI-controlled equipment would purify it with little human intervention. From the mine face to the product that is ready for the market, the whole value chain becomes efficient, smart, and flexible.
- **The Fully Autonomous Mine**: In the end, the future mine will function similarly to a living thing: it will sense, think, adapt, and act on its own. Each robot, drone, sensor, and equipment will be a part of an autonomous mining network that continuously maximizes sustainability, productivity, and safety. Human intervention in such a setting shifts from being operational to strategic, concentrating on innovation, decision-making, and supervision.

In summary, the emergence of autonomous mining is about building intelligent mining ecosystems where humans, artificial intelligence, and robotics work together to unlock resources more effectively, safely, and sustainably than ever before. It is not only about machines taking the place of people.

5.7. Real-Time Human-Machine Cooperation

The smooth interaction between intelligent systems and humans is the real power of AI and IoT in underground mining, not merely the automation of machinery. In this new mining paradigm, decision-making, efficiency, and safety take place in real time inside an integrated ecosystem made up of operators, equipment, and digital platforms.

5.7.1. Real-Time Coordination

Every piece of machinery and every employee in a modern underground mine is tracked and monitored via a computerized network, making them resemble living things.

- Autonomous loaders immediately reroute a congested haulage tunnel, preserving efficiency and preventing delays;
- AI-driven coordination platforms continuously assess the position, speed, and operating status of trucks, drills, and loaders to guarantee that machine traffic flows without collisions or bottlenecks.
- AR helmets and wearable technology provide real-time alerts to human workers, alerting them when a path ahead is blocked, when machinery is coming, or when they need to intervene for a particular activity.

Underground, where people and machines work together harmoniously rather than in competition for space, this dynamic flow of information produces a symphony of movement.

5.7.2. Predictive Support

Anticipatory AI systems that serve as digital guardians improve human productivity and safety.

- Miners' heart rates, levels of exhaustion, and exposure to environmental dangers like dust, noise, or poisonous gasses are all continuously monitored by wearable sensors. Supervisors are immediately alerted and employees receive alerts if thresholds are exceeded.
- At the same time, AI forecasts when vital gear will require maintenance. Systems proactively plan rerouting, part replacements, or temporary shutdowns rather than waiting for equipment to malfunction, reducing human risk and preventing expensive delays. By ensuring that both people and machinery stay in optimal condition, this predictive help significantly increases operational uptime.

5.7.3. Dynamic Tasks Allocation

Adaptive workforce management is introduced underground by AI and digital twins:

- Digital twins evaluate workloads and determine whether a task is better performed by an autonomous machine or a human by simulating activities in real time.
- Human exposure to risk is decreased when high-risk tasks, like blasting preparation or working in unstable tunnels, are frequently assigned to robotic systems.
- On the other hand, duties that call for sophisticated judgment or inventiveness—like resolving unforeseen geological anomalies—remain within human control.

By ensuring that each duty is carried out by the most competent agent—human or machine—at the appropriate moment, this dynamic assignment of responsibilities maximizes efficiency and safety.

5.7.4. Emergency Response and Safety Procedures

Instantaneous, AI-driven response systems are revolutionizing underground mining safety:

- Gas leaks, structural flaws, rock collapses, and unexpected flooding are all detected by networks of Internet of Things sensors.
- AI algorithms instantly start emergency procedures, such locking down susceptible locations, starting ventilation systems, or rerouting autonomous transporters away from dangerous areas.
- Automated evacuation orders are given to human personnel, who are directed to the closest safe area by AR headsets.
- In an emergency, machines are also coordinated; robots may mobilize to secure infrastructure, while drones may be sent out to evaluate potential threats.

As a result, there is a human-machine emergency response system that is well-coordinated, which significantly lowers the number of casualties, avoids operational chaos, and guarantees disaster resilience.

5.8. The Next Development: Humans and Machines Symbiosis

Underground mining will transition from collaboration to full symbiosis as AI, IoT, and sophisticated interfaces develop. In this scenario, humans and machines will share knowledge, awareness, and even make decisions in real time:

5.8.1. Cognitive Collaboration and Neural Interfaces

Brain-Computer Interfaces (BCIs), which enable operators to communicate with autonomous systems solely through thinking, will be used in the next generation of mining equipment:

- Without clicking buttons or giving spoken instructions, a miner might reroute an autonomous loader by simply imagining a path.
- Neural link-connected AI co-pilots are able to predict an operator's intentions and make recommendations or carry out standard adjustments on their own.
- By streamlining intricate processes and significantly lowering reaction times in emergency situations, this degree of integration fosters immediate synergy between machine execution and human intuition.

5.8.2. Augmented Reality Overlays

In the future, systems will project *immersive augmented reality (AR) overlays* straight into a miner's field of view, replacing the notifications that wearables currently offer:

- AR will also offer *step-by-step instructions for equipment*

maintenance, enabling less experienced workers to carry out complex tasks safely under AI supervision.
- Workers will see dynamic hazard maps projected onto tunnel walls, highlighting unstable rock zones, gas concentrations, or restricted machinery paths.
- Remotely connected supervisors are able to "see through the eyes" of subterranean workers, providing guidance and support in real time.

5.8.3. AI Assistants for Each Employee

Each miner will have a digital assistant, or *AI Co-pilot*, built into their neural interface or wearable:

- The co-pilot provides context-aware advice by keeping an eye on environmental data, equipment health, and human performance.
- For instance, the AI co-pilot might advocate micro-breaks or work reassignment if weariness symptoms increase.
- The AI co-pilot can automatically rectify small mistakes in crucial activities, such as precision drilling, guaranteeing accuracy without interfering with the workflow of the human operator.

5.8.4. Networks of Collective Intelligence

Mines will create collective intelligence systems as a result of all people and machines being connected by IoT and AI:

- All decisions, equipment adjustments, and hazard detections are instantaneously sent throughout the network.
- All workers' AR systems instantly update if one miner comes upon an unsafe tunnel, and machines adjust their routing

accordingly.

- This raises safety and productivity standards throughout the entire operation by producing a hive-mind effect, in which one person's experience becomes the knowledge of all.

5.8.5. Systems of Emotional and Cognitive Safety

In addition to tracking physical health, future wearables and BCIs will also keep an eye on cognitive and emotional states:

- AI co-pilots may modify workloads, pause machine engagements, or initiate supportive interventions in response to stress, distraction, or cognitive overload.
- This makes mining a more sustainable and compassionate industry by guaranteeing that people are not only physically safe but also psychologically resilient.

5.9. Redefining Mining's Social Contract

AI and IoT are not replacing people in this changing environment; rather, they are redefining the core function of human labour in mining.

Risky, repetitive jobs give way to strategic oversight, innovation, and augmentation. While machines manage the risky activities, heavy lifting, and precise drilling, workers take on the roles of commander, analyser, and innovator.

In addition to increasing productivity, this new human-machine connection reframes mining as a sector where advancement is centred on safety, intelligence, and human dignity.

5.10. Case Study: Australia's Autonomous Gold Mine

One of the most highly sophisticated gold mines in the world is revolutionizing 21st-century mining in the centre of Western Australia. This website has established itself as a world standard for combining AI-powered coordination, autonomous systems, and human-machine cooperation.

- **Scalable Autonomous Haulage**: Currently, autonomous trucks and loaders handle around 70% of all underground haulage activities in this mine. Without the need for human drivers, these vehicles can coordinate with other robots, navigate tight tunnels, and optimize their routes in real time. As a result, there are much fewer accidents, less fuel usage, and continuous, round-the-clock hauling with fewer disruptions.
- **AI-Powered Complex Operation Orchestration**: Behind the scenes, the mine's digital brain is a sophisticated AI coordination platform. Drills, loaders, drones, inspection robots, and human operators are among the more than 150 interconnected assets it concurrently oversees. The technology integrates human decision-making when necessary while maintaining the fluidity of machine-to-machine (M2M) activities by processing thousands of data points each second.
- **Increases in Productivity and Safety**: Due to 24/7 operations and improved workflows, the mine has seen a 30% boost in overall productivity since introduction. At the same time, there has been a 45% decrease in near-miss accidents, demonstrating how automation may greatly improve worker safety when properly coordinated. Instead of performing

high-risk or repetitive operations, human workers are now largely in charge of oversight, system administration, and strategic decision-making.

- **Real-Time Communication as a Backbone**: The success of the mine depends not only on the machinery but also on the backbone of real-time communication that links each operator, sensor, and autonomous system. AI-driven communication networks quickly synchronize all actions, whether a miner makes a strategic call, a drone identifies a structural hazard, or an autonomous loader reroutes in the middle of an operation. Conflicts are avoided, downtime is reduced, and human judgment is smoothly incorporated into automated procedures thanks to this connection.
- **The Human Factor**: Most importantly, the mine shows that human knowledge is still essential in highly automated settings. Operators, engineers, and geologists offer oversight and insights that machines cannot match. These ground-breaking outcomes come from the fusion of human creativity and machine accuracy, made possible by AI coordination.

Important Takeaway: The gold mine in Western Australia demonstrates that creating a symbiotic environment in which humans, artificial intelligence, and self-governing machines collaborate is more important for mining in the future than replacing people with robots. A strong AI-driven communication infrastructure that integrates people and technology into a single intelligent system is necessary for true operational change, which calls for more than just sophisticated gear.

5.11. Important Technologies That Make Coordination Possible

Today's safe and effective mining operations rely on a variety of cutting-edge technologies that allow human operators and autonomous systems to collaborate in real time. Every choice, whether made by humans or computers, is well-informed, accurate, and coordinated thanks to these technologies, which build an intelligent, networked ecosystem beneath the surface.

5.11.1. Edge Computing

Instead of sending data to a remote cloud server, edge computing enables local processing at the source. For underground mining, this offers a number of significant benefits:

- **Reduced Latency**: By making snap decisions without awaiting distant calculation, autonomous cars, drones, and robotic systems lower the chance of collisions or operational delays.
- **Increased Reliability:** Edge computing guarantees that computers continue to operate effectively even in locations with sporadic network connectivity.
- **Data Prioritization**: This technique lowers bandwidth needs while preserving real-time intelligence by sending just the most important data to central servers.

  Edge computing serves as the intelligent mine's nervous system by decentralizing processing, allowing for quick reactions to shifting conditions below ground.

5.11.2. AI-Driven Task Scheduling

AI-powered task scheduling systems coordinate human and machine activity, guaranteeing that every process is optimized for effectiveness and security:

- **Enhanced Throughput**: Algorithms constantly assess workloads and, based on availability, capability, and safety, allocate jobs to the best agent, whether it be an autonomous system or a human.
- **Reduced Idle Time**: When bottlenecks arise, staff and machines are immediately redistributed to ensure that no resource is left idle while operations proceed.
- **Dynamic Adaptation:** Scheduling adjusts in real time to unforeseen circumstances, like equipment failures, environmental dangers, or changes in the priorities of production.

Because of the AI coordination, mining is no longer a reactive operation but rather a predictive, proactive workflow where each task is carried out precisely and with the least amount of waste.

5.11.3. Cobots (Collaborative Robotics) Cobots are robots made to securely work alongside people by fusing human judgment with endurance and accuracy:

- **Precision Tasks:** Cobots carry out tasks requiring a high degree of accuracy, like sorting materials, drilling, and assembling equipment.
- **Safety Integration**: Cobots that are outfitted with sensors and artificial intelligence (AI) vision are able to recognize human presence and automatically modify their actions to prevent collisions.
- **Human Oversight**: Miners concentrate on intricate decision-

making, troubleshooting, and quality control, whereas Cobots manage repetitive or hazardous jobs.

Cobots reduce exposure to repetitive or hazardous tasks while increasing productivity by fusing human intellect with robotic efficiency.

5.11.4. Interfaces for Augmented Reality (AR)

Miners and supervisors may see underground operations in real time thanks to AR interfaces, which speeds up and improves decision-making:

- **Operational Awareness**: In a miner's range of vision, augmented reality overlays can provide tunnel layouts, machine status, hazard zones, and even predictive maintenance notifications.
- **Guided Interventions**: Employees carrying out maintenance or inspection duties are given detailed AR instructions, which lowers errors and expedites intricate processes.
- **Remote Collaboration**: Using augmented reality streams, supervisors can watch operations from a distance and give real-time guidance or support without having to be in dangerous areas.

With AR, the mine is transformed into a dynamic, interactive workspace where all operators have immediate access to the data they need to make wise decisions.

These technologies work together to create a single, intelligent system that allows humans and machines to conduct coordinated underground activities. AI streamlines processes, Cobots securely carry out precise tasks, edge computing guarantees quick reactions, and augmented reality gives people real-time awareness. These technologies work together to make modern autonomous mining safe and effective.

5.12. Completely Integrated Human-Machine Ecosystems: The Future

Beyond basic automation, the next phase of underground mining will involve *fully cognitive, self-organizing ecosystems* in which robots and humans work together in real time. These mines will behave as intelligent, adaptive networks that can learn, optimize, and react on their own, rather than as groups of separate machines and labourers.

5.12.1. Scalable Autonomous Coordination

- *Without direct human interference*, machines will coordinate with one another as well as with humans, dynamically modifying tasks, routes, and schedules in response to real-time data.
- A *coordinated fleet* of autonomous loaders, drills, drones, and inspection robots will react immediately to modifications in tunnel conditions, ore availability, or hazard identification.
- While regular operations, danger mitigation, and equipment optimization are managed independently, human operators take on the role of *strategic supervisors,* making high-level choices.

5.12.2. Adaptive Intelligence And Continuous Learning

- Every action taken in the mine, whether by a machine or a human, is recorded, examined, and used to enhance performance in the future.
- Artificial intelligence (AI) systems will recognize trends in human behaviour, environmental changes, and equipment usage to build *predictive models* that continuously improve operational efficiency, productivity, and safety.

- The ecosystem develops into a *self-optimizing* mine over time, making decisions more quickly, accurately, and securely than any system that is only controlled by humans.

5.12.3. Networked Multi-Mine Operations:

No longer will autonomous coordination be limited to a single location. Multiple mines in different regions will be connected by intelligent networks, enabling operations to pool resources, reroute machinery, and maximize output as a group:

- Centralized AI platforms will offer a pan-regional perspective of mining operations, distributing machines across sites for optimal efficiency, monitoring environmental effect, and balancing workloads.
- These interconnected ecosystems will change the mining sector from discrete operations into hyper-connected, data-driven networks that can adjust to unanticipated disturbances or the demands of regional markets.

5.12.4. Combined Results: Profitability, Safety, and Productivity

- In these highly interconnected human-machine ecosystems, safety, profitability, and productivity are not separate goals but rather reinforce one another.
- Why Safer operations result in fewer incidents and downtime, which immediately increases output.
- Predictive coordination and ongoing AI learning improve operational effectiveness and resource usage, which raises profitability.
- Machines perform repetitive, high-precision, or high-risk

jobs, while human expertise is deliberately used where it delivers the greatest value.

Diagram: The Digital Nervous System of the Mine

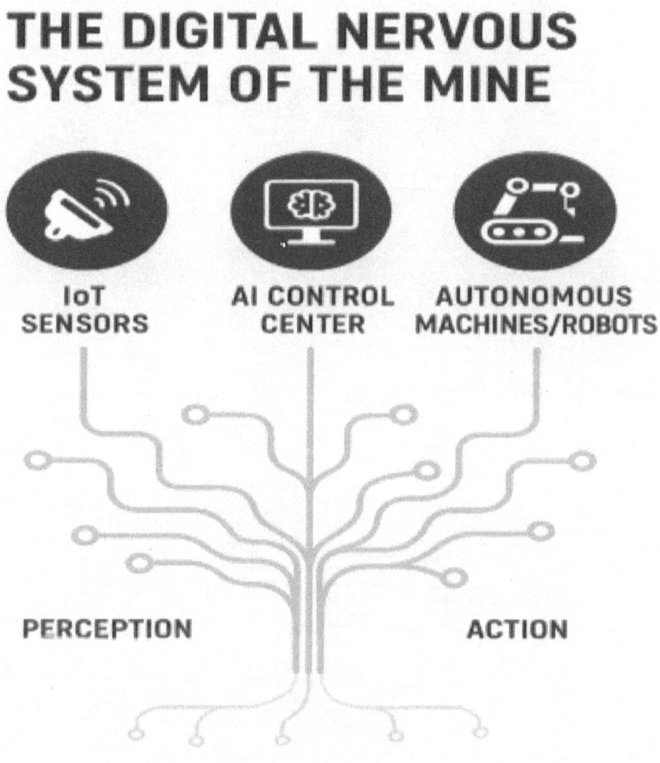

Figure 6: The Digital Nervous System of the Mine

As a result, the fully integrated mine of the future is not only autonomous but also intelligent, adaptive, and symbiotic, functioning as a single, cohesive unit with both humans and machines. Every choice, action, and invention made in such a system concurrently advances economic performance, safety, and efficiency, ushering in a new era of underground mining.

Diagram: The Future Fully Integrated Human-Machine Ecosystem

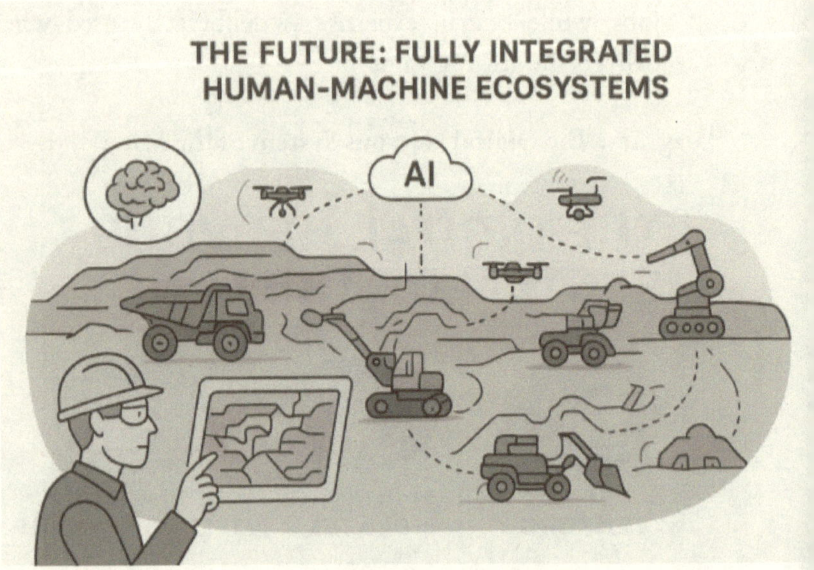

Figure 7: The Future Fully Integrated Human-Machine Ecosystem

Chapter 6: Safety by Design: AI-Powered Emergency Response, Evacuation, and Hazard Detection

Overview

Safety is essential to underground mining since it is the cornerstone of output, operational stability, and worker well-being. Although somewhat successful, traditional safety measures are reactive by nature since they depend on post-event interventions, manual inspections, static alarms, and human monitoring. Even though these techniques have been used historically, they are not very good at foreseeing complicated risks, particularly in dynamic subterranean environments where conditions might change suddenly.

By turning mine safety from a reactive requirement into a proactive, predictive, and adaptable system, AI and IoT technologies have the potential to completely change the industry. By integrating sensor networks, machine learning algorithms, and automated decision-making, mines can achieve "Safety by Design," embedding risk mitigation into every aspect of operations. The following sections explore how this is achieved in practice.

6.1. Predictive Hazard Detection

AI-powered hazard detection is transforming mine safety by leveraging massive volumes of real-time data from IoT sensors deployed across underground and surface operations. These sensors continuously monitor critical environmental, structural, and operational parameters, enabling the system to anticipate dangerous events before they occur. Key areas of focus include:

- **Monitoring of Air Quality**: Sensors identify dangerous gases including carbon monoxide, methane, and other flammable or toxic materials. High levels of dangerous gasses and oxygen depletion are detected right away. In order to predict possible build-ups and enable proactive ventilation modifications or evacuation alarms, artificial intelligence algorithms examine trends and patterns in gas levels.
- **Structural Integrity Surveillance**: Rock movements, ground tension, and vibrations are all potential hazards in mines, which are naturally dynamic environments. IoT-enabled structural sensors monitor ground movement, micro-fractures, and stress spots. By identifying minute departures from typical patterns, machine learning models can anticipate collapses or rock falls before they become serious occurrences.
- **Equipment Condition Monitoring**: Every piece of mining equipment, including vehicles, drills, and conveyors, is regularly checked for abnormalities in performance. Temperature, vibration, pressure, and operating efficiency data are all gathered via sensors. Artificial intelligence (AI) systems identify early indicators of mechanical wear or failure, such as hydraulic leaks or overheating bearings, which could result in explosions, fires, or other operational dangers.
- **Worker Proximity and Movement Tracking:** Wearable

technology and location-tracking sensors further improve worker safety. Real-time worker position monitoring by AI systems detects possible equipment collisions and hazardous proximity to unstable locations. Instant alerts can be issued to control rooms and employees, preventing accidents or entrapment.

- **Equipment Condition Monitoring**: Every piece of mining equipment, including vehicles, drills, and conveyors, is regularly checked for abnormalities in performance. Temperature, vibration, pressure, and operating efficiency data are all gathered via sensors. Artificial intelligence (AI) systems identify early indicators of mechanical wear or failure, such as hydraulic leaks or overheating bearings, which could result in explosions, fires, or other operational dangers.
- **Worker Proximity and Movement Tracking:** Wearable technology and location-tracking sensors further improve worker safety. Real-time worker position monitoring by AI systems detects possible equipment collisions and hazardous proximity to unstable locations. Instant alerts can be issued to control rooms and employees, preventing accidents or entrapment.
- **Predictive Analytics for Proactive Safety:** By combining all of these data sources, machine learning models are able to spot trends and abnormalities that human observers would miss. For example, minute variations in the behaviour of equipment may foretell a severe breakdown, or faint vibrations in a tunnel may signal an impending collapse. Through proactive interventions made possible by AI's recognition of these early warning indicators, miners can evacuate or take care of risks before they become more serious.
- **Result:** A predictive safety network that is always learning is the end product. The AI system gains accuracy over time

as a result of each incidence, sensor reading, and operational event. Mines drastically cut down on mishaps, equipment damage, and downtime by switching from reactive safety measures to proactive hazard prevention.

6.2. Intelligent Evacuation Systems

Conventional mine evacuation methods frequently depend on human guidance, static signage, and set exit pathways. Although these techniques work well in the best-case scenario, they may not be sufficient in situations that change quickly, including tunnel collapses, gas leaks, or fires. These constraints are overcome by AI-driven intelligent evacuation systems, which maximize speed and safety by dynamically adjusting to current conditions.

- **Flexible Route Scheduling**: To identify the safest and quickest evacuation routes, AI algorithms continuously evaluate sensor data, including heat maps, smoke density, poisonous gas levels, and structural integrity indicators. For instance, the technology instantly recalculates alternate paths to avoid risk zones if a rock fall compromises a primary tunnel. In order to give workers in underground mines real-time route changes, businesses such as **Sandvik and Caterpillar** are actually creating autonomous systems that combine mine maps with sensor networks.
- **Personalized Alerts and Guidance:** Miners can get personalized instructions thanks to wearable technologies, such as smart helmets, augmented reality glasses, and wrist devices. Workers are guided along paths that avoid hazards by turn-by-turn guidance superimposed on their surroundings in place of general evacuation orders. Trials of AR-enabled evacuation systems in Australian coal mines have shown a 30–40% decrease in egress times during simulated crises, underscoring the importance of tailored instruction.
- **Automated Environmental Control:** To enable safe evacuations, AI systems can actively modify mine infrastructure. For example:

  - Lighting can dynamically illuminate safe paths while designating obstructed or unsafe places; ventilation systems can be changed to divert smoke and poisonous gasses away from evacuation routes.
  - To limit risks and keep miners out of hazardous areas, emergency doors and barriers can be programmed to open or close automatically.
  - One prominent example is the Vale iron ore mines in Brazil, where AI monitoring combined with automated ventilation and barrier control systems has increased the effectiveness of emergency evacuations and reduced exposure to dangerous situations.
- **Continuous Real-Time Analysis:** Intelligent evacuation systems update risk assessments and evacuation plans on a regular basis by combining data from environmental sensors, equipment monitors, and personnel tracking devices.

This guarantees that the system rapidly recalibrates routes and directions in the event of a sudden gas leak or equipment explosion. Compared to static evacuation plans, this method actually improves overall safety, lessens traffic, and avoids panic.

- **Result:** The paradigm shift from reactive to proactive mine safety is shown by intelligent evacuation systems. Mines can significantly lower the number of fatalities, injuries, and operational disruptions by integrating wearable technology, AI analytics, and automated environmental controls to make evacuations safer, quicker, and customized to changing threats.

6.3. Automated Emergency Response

AI-driven automated emergency response raises the bar for mine safety by coordinating rescue and containment operations with previously unheard-of speed, accuracy, and intelligence, while predictive hazard detection and intelligent evacuation enhance prevention and egress. These devices can significantly lower casualties and operational losses by reducing human exposure to hazard and improving decision-making.

6.3.1. Autonomous Rescue Robots and Drones:

Drones and robotic units with AI capabilities are being used more frequently to traverse dangerous areas that are unsafe or inaccessible to human rescuers. In order to detect heat, gas, structural instability, and the existence of trapped individuals, these devices are outfitted with sensors, thermal cameras, and environmental monitors.

For example:

- Autonomous drones have been tried for search-and-rescue exercises in **Australian underground coal mines**, effectively locating fictitious victims in collapsed tunnels and sending real-time location and health data back to control centres.
- Before human crews can safely get at the scene, robotic rescue vehicles created by firms like **Exyn Technologies and Clearpath Robotics** can provide oxygen supplies, first aid kits, and communication devices to trapped miners.

6.3.2. Integrated Communication Networks:

Standard communication routes are frequently compromised in emergency situations, making rescuers unaware of the facts on the ground. Strong, flexible communication networks between subterranean staff, surface command centres, and outside emergency responders are made possible by AI. These networks can seamlessly transition between wired, wireless, and mesh systems while preserving data flow. Artificial intelligence (AI)-enabled communication systems were able to maintain data transmission during simulated tunnel collapses at mines like **Vale's operations in Brazil**, enabling surface crews to coordinate reactions even in locations with historically poor signal coverage.

6.3.3. Incident Simulation and Decision Support: AI models are capable of simulating a variety of emergency situations, such as fires, gas leaks, and structural collapses, which enables decision-makers to instantly prepare the best containment tactics. AI offers practical suggestions on resource deployment, evacuation prioritizing, and rescue sequencing by evaluating variables including personnel locations, hazard severity, and available resources. An example from **Rio Tinto's mining operations** shows how AI-powered simulation improved emergency drills, cutting response times by more than 25% and enhancing team collaboration.

6.3.4. Real-Time Situational Awareness: Data from sensors, drones, robots, and personnel trackers is continuously integrated by automated emergency response systems. This creates a thorough situational map of the mine that shows the locations of survivors, safe areas, and risks.

By acting strategically instead of reactively, rescue teams can lower the likelihood of subsequent mishaps.

6.3.5. Result: Automated emergency response systems transform mine safety by fusing autonomous robotics, robust communication networks, and AI-powered decision assistance. Life is saved, evacuation and rescue operations are optimized, and human responders are protected from severe danger. These solutions enable mines to develop a completely integrated, intelligent emergency ecosystem, going beyond reactive safety procedures.

6.4. Continuous Learning and Safety Enhancement

AI-driven mine safety solutions are distinguished by their capacity for continuous learning, adaptation, and improvement. AI systems establish a constantly changing safety ecosystem that foresees dangers, improves processes, and lowers accidents through continuous intelligence, in contrast to traditional safety protocols, which are static and reactive.

6.4.1. Data-Driven Insights: Every bit of information is recorded and examined, from small irregularities to significant occurrences. AI models are fed by near-misses, equipment failures, environmental changes, and even patterns of human behaviour. AI is able to anticipate risks that would otherwise go undetected by spotting patterns and correlations.

For instance:

- To detect patterns of mechanical stress in **BHP's underground mining** operations, AI algorithms examine sensor data from regular equipment inspections and near-miss reports. This results in proactive repair plans and a notable decrease in accidents involving equipment.

6.4.2. Adaptive Safety Protocols: Using both previous and current data, AI systems continuously improve emergency protocols. As the system learns from each incident, hazard warnings, ventilation modifications, and evacuation routes change. Instead of depending just on established regulations, this flexibility guarantees that safety precautions are always in line with the circumstances at hand.

- One example is the use of AI-driven adaptive protocols by **Anglo American Platinum,** which enable safer and quicker egress during drills and real emergencies by updating evacuation and emergency response plans in real time depending on worker movements and mine conditions.

6.4.3. Feedback Loops from Staff and Equipment: Feedback from a variety of sources is necessary for safety optimization:

- **Behaviour of Miners**: Wearable technology monitors worker speed, movement, and compliance with safety regulations. To enhance training and prediction models, deviations are noted and examined.
- **Equipment Performance**: Sensors keep an eye on the condition of the machinery, warning AI systems of possible risks from wear, overheating, or malfunctions.
- **Environmental Trends**: Risk assessment models are informed by ongoing observations of vibrations, air quality, and structural stability. A self-improving system is produced by this input, as the AI determines which interventions work best and modifies suggestions for subsequent operations.

In actuality, miners employing these systems report quantifiable increases in incident prevention, decreased downtime, and safety compliance.

6.4.4. Transition a Living Safety Ecosystem: Mines go from static, compliance-driven safety approaches to intelligent, anticipatory safety ecosystems through ongoing learning. Emergency protocols are dynamically improved, hazards are identified and reduced before they materialize, and the mining environment is made safer for both people and machinery. In the long run, this strategy not only saves lives but also boosts company safety culture, lowers insurance costs, and increases operational efficiency.

- **Case Study:** As an illustration, consider **Rio Tinto's "Autonomous Safety System,"** which continuously modifies safety precautions by combining real-time sensor data with AI learning loops. It has been demonstrated that the system lowers incident rates and boosts employee confidence, illustrating how ongoing learning directly benefits operations and people.

6.4.5. Strategic Benefits of AI-Driven Safety: Adopting AI and IoT in mine safety offers quantifiable operational, financial, and human benefits; it is much more than just a compliance exercise. Organizations build robust, effective, and sustainable work environments by including safety into the very foundation of mining operations.

- **Decreased Downtime**:

Conventional mining operations are susceptible to unforeseen events that might stop output for hours or even days, such as collapses, gas leaks, or equipment breakdowns. Predictive danger detection powered by AI detects threats before they become more serious, allowing for preventative measures.

- **Example:** BHP's underground operations have seen a 20% reduction in unscheduled downtime due to the use of predictive analytics and IoT monitoring in preventing equipment-related issues. This optimizes labour planning and resource use in addition to maintaining output on schedule.

- **Reduced Insurance Costs:** Liability exposure and insurance costs are directly impacted by workplace accidents and safety infractions. Because AI-driven safety systems cut incident rates, insurers are willing to slash premiums for mines that use them.
    - **Case Study**: According to Anglo American, including AI monitoring and predictive safety procedures resulted in a quantifiable decrease in workplace injuries, which in turn led to fewer insurance claims and advantageous coverage cost modifications.
- **Increased Workforce Confidence:** Operational effectiveness depends on employee confidence and morale. When miners work in AI-monitored workplaces, they may rest easy knowing that dangers are constantly being watched over and that emergency reaction and evacuation plans are proactive rather than reactive.
    - **Example:** Mines that use wearable safety technologies and real-time monitoring, like **Vale's facilities in Brazil**, report improved worker satisfaction and lower stress levels, which boost output and decrease turnover.
- **Regulatory Alignment and Compliance**: Safety requirements and mining laws are always changing. By ensuring that operations maintain compliance with air quality, structural integrity, and occupational safety regulations, real-time AI monitoring helps to prevent fines or shutdowns.

  - **Example:** AI-enabled reporting systems in Australia and the European Union streamline audits and show conformity with national and international rules by automatically creating compliance records, highlighting irregularities, and recommending corrective actions.
- **Sustainability and Operational Resilience:** Mines can protect employees, equipment, production schedules, and company reputation by integrating AI-driven safety into the operating fabric. Predictive safety minimizes disruptions, maximizes the use of available resources, and promotes a continuous improvement culture. These solutions gradually turn safety from a reactive duty into a strategic advantage, promoting long-term viability and financial success.
  - Higher operational resilience and increased stakeholder confidence are reported by mines implementing AI safety platforms, such as **Rio Tinto's autonomous and integrated safety systems**.
- **Result:** Beyond compliance, AI-driven safety offers real economic benefits. Mines obtain a competitive advantage while preserving human life by lowering insurance premiums and downtime, increasing employee satisfaction, and guaranteeing regulatory compliance. Safety no longer serves only as a legal necessity but also as a catalyst for effectiveness, adaptability, and sustainable growth.

In conclusion, AI and IoT are revolutionizing underground mine safety by moving away from reactive measures and toward automated, intelligent, and proactive defense. "Safety by Design" has evolved from a theoretical idea to a data-driven, practical strategy that protects lives, streamlines operations, and increases the resilience of mining ecosystems. Every choice, sensor reading, and machine action in the age of hyperconnected mines adds to a constantly changing safety intelligence network.

6.5. Vision 2035: Autonomous Ecosystems in Underground Mining

By 2035, underground mine safety will be a completely autonomous, self-optimizing ecosystem thanks to AI and IoT, rather than a reactive requirement. In this vision:

6.5.1. Completely Autonomous Hazard Anticipation:

- AI systems will continuously track human activity, equipment performance, atmospheric variations, and geological conditions in real-time.
- Cutting-edge predictive models will spot possible risks like gas pockets forming, tunnel micro fractures, or equipment failures seconds before they materialize.
- Risk mitigation will be immediate, quick, and undetectable; it will stop accidents before people ever notice them. Mines will no longer wait for alarms to sound.

6.5.2. Self-Directed Evacuation:

- AI-enabled wearable technology will direct miners along dynamically optimized escape routes that are always adapting to changing circumstances.
- Drones and autonomous cars will move miners out of unsafe areas, allowing for evacuation even in areas that are too risky for people to travel through.
- To guarantee the safest possible egress, AI will autonomously operate ventilation systems, safety barriers, and emergency exits in real-time. AI-controlled in real-time to guarantee the safest exit.

6.5.3. AI-Orchestrated Rescue Operations:

- Should an incident occur, autonomous robots and drones will immediately go into action to locate, stabilize, and remove any impacted persons?
- AI will communicate updates directly to command centres and miners, maintaining situational awareness and lowering uncertainty in crisis management.
- AI will simulate and carry out rescue strategies in real-time, coordinating multiple devices, personnel, and resources with precision that human teams cannot match.

6.5.4. Continuous Self-Learning and Adaptation

- An ever-evolving AI model will be fed by each near-miss, equipment anomaly, and environmental change.
- Mines will operate as live safety networks, capable of predicting new hazard patterns, autonomously updating protocols, and training human teams through simulations based on historical and data.
- AI will manage the operational implementation of safety, while human oversight will transition from risk monitoring to strategic direction.

6.5.5. Result: Underground Mining Without Incidents

- Underground mining will be safer than ever by 2035 thanks to AI-driven hazard detection, autonomous evacuation, and automated emergency response.
- In order to maximize efficiency and safeguard the workforce, mining operations will strive for a condition of almost zero accidents.
- From design and operations to real-time response and

ongoing improvement, safety will no longer be an afterthought; rather, it will be integrated into every aspect of the mine's ecosystem.

According to this view, AI will be the ultimate protector of mining, fostering an atmosphere in which machine intelligence and human creativity work together harmoniously to achieve hitherto unheard-of levels of resilience, efficiency, and safety.

6.6. The Problem of Underground Safety

The conditions for underground mining are inherently dangerous and uncertain. Geological instability, heavy machinery, and cramped quarters combine to produce an environment where risks are not only likely but unavoidable. Underground mining, in contrast to surface activities, reduces visibility, obstructs escape routes, and intensifies the effects of even small errors. The main dangers consist of:

6.6.1. Explosions and Gas Leaks

- Conventional detection depends on stationary sensors or sporadic manual checks, which leave gaps in coverage and reaction time;
- Mines can accumulate toxic and flammable gases like hydrogen sulphide, carbon monoxide, and methane;
- Even low concentrations can be harmful to health; and
- Sudden build-ups can result in devastating explosions. Continuous monitoring powered by AI can identify minute variations, allowing for preventative measures to be taken before concentrations hit risky thresholds.

6.6.2. Floods And Water Ingress:

- Unexpected floods might be caused by old tunnels, underground water sources, or malfunctioning pumps, endangering both people and property.
- Water intrusion can disrupt operations, short circuit electrical systems, and jeopardize tunnel structure.
- In an emergency, prompt identification and reaction are essential; if alarms are delayed, miners may become trapped or equipment damage may be irreparable. By analysing

geological and hydrological data, advanced AI prediction models can predict water input before it reaches dangerous levels, enabling automated sealing or prompt evacuations.

6.6.3. Structural Instability:

- Rock falls, tunnel subsidence, and roof collapses are recurring hazards brought on by seismic activity, excavation vibrations, or geological flaws.
- Because these occurrences are frequently abrupt, there is little time for manual intervention.
- By utilizing IoT sensors in conjunction with AI analytics, structural monitoring may detect stress patterns and early warning indicators, providing miners and supervisors with crucial seconds to take action, averting injuries and devastating collapses.

6.6.4. Equipment Failure:

- Although heavy equipment like drills, loaders, and conveyor systems are essential to operations, wear, lack of maintenance, or environmental factors can cause them to malfunction.
- Subterranean equipment failure increases the risk of other hazards by causing production disruptions, blocking escape routes, and injuring personnel.
- AI-powered predictive maintenance makes sure that equipment is maintained before it breaks down, thereby lowering the risk of mishaps.

6.6.5. The Price of Postponement

- Milliseconds count in subterranean mines. A simple problem can become life-threatening due to a missed sensor reading, a delayed warning, or a poorly planned evacuation.

- Human monitoring is insufficient on its own because susceptibility is increased by weariness, poor vision, and communication lags. Artificial intelligence (AI)-powered safety systems can bridge the gap between hazard detection and action by processing complex streams of data in real-time, identifying risks more quickly than humans, and coordinating prompt responses.

In summary, gases, water, structural instability, and equipment hazards all combine to create a dynamic environment where traditional safety precautions frequently fail, making underground safety a multifaceted problem. Integrating AI and IoT is crucial in this situation; it is not an alternative. Intelligent safety solutions change the underground mine from a high-risk area into a regulated, proactive setting where risks are identified, reduced, and controlled before they become catastrophes by offering real-time awareness, predictive insights, and automatic responses.

6.7. Matrix of Underground Hazard Risks

It is crucial to assess each hazard according to its likelihood and potential severity in order to comprehend and successfully manage underground mining risks. A risk matrix offers a visual framework that shows which hazards need to be addressed right away and how AI-driven interventions can lessen them.

6.7.1. How AI Modifies the Risk Environment

- Conventional safety methods, which depend on inspections and human judgment, treat hazards as static dangers.
- AI changes this by offering dynamic risk assessment, in which the likelihood and seriousness of each threat are continuously updated by each sensor input, environmental change, and operational activity.
- Rather than responding after an incident happens, the outcome is a living risk matrix that automatically prioritizes threats and deploys resources proactively.

6.7.2. Visualising AI-Enhanced Risk Management

Imagine the central command of the mine having a digital control dashboard:

- Green indicates low risk, yellow indicates moderate risk, and red indicates serious risk.
- AI forecasts made in real time continuously modify these risk levels.
- Preventive action is taken milliseconds before danger rises thanks to automated alarms and intervention processes that are activated the moment a hazard reaches a critical level.

In conclusion, underground mining transitions from reactive safety procedures to proactive hazard management by fusing an IoT intelligence and AI with a risk matrix. By transforming high-risk areas into intelligent, self-monitoring ecosystems, mines are able to predict hazards, maximize reactions, and save human lives in a complex, quickly changing environment.

6.8. AI-Powered Hazard Identification

Hazards in underground mining can materialize in a matter of seconds, and prompt, well-informed action frequently makes the difference between safety and disaster. The limitations of traditional monitoring systems include their passive data collection and reliance on human interpretation, which results in gaps in accuracy and speed. This paradigm shift is brought about by AI, which creates a predictive, self-correcting safety network and turns raw sensor data into usable, real-time insights.

6.8.1. Real-Time Environmental Monitoring

- IoT sensors placed throughout the mine continuously track temperature, humidity, seismic activity, vibrations, and gas concentrations (methane, carbon monoxide, and oxygen levels). These data streams are analysed by AI algorithms to find anomalies, differentiating between normal variations and risky patterns that indicate new dangers.
- To ensure preventive action before circumstances reach dangerous levels, for instance, an abrupt surge in methane in a tunnel that is rarely visited would immediately result in an AI-generated alert, even if the reading is below conventional alarm limits.
- By integrating wearable technology, miners can get location-specific, tailored warnings that improve situational awareness and speed up decision-making.

6.8.2. Predictive Risk Analytics

- AI makes predictive risk analytics possible, which identifies possible risks before they become apparent, in addition to real-time monitoring.

- To predict disasters like tunnel collapses, gas leaks, or water intrusion, machine learning algorithms analyse past sensor data, operational records, geological studies, and seismic trends.
- **Example:** AI can identify patterns of stress accumulation in the rock ceiling by examining long-term vibration data in a particular mine area. The technology proactively sends out evacuation notifications and suggests reinforcement actions if these patterns coincide with past signatures linked to collapses.
- By optimizing maintenance plans, resource allocation, and staff placement, predictive analytics also helps miners minimize operating disruptions and lower exposure to possible risks.

6.8.3. Automated Safety Inspections of Equipment

- Although heavy equipment is necessary for mining operations, malfunctioning equipment can be dangerous. AI tackles this by means of autonomous intervention and ongoing monitoring.
- Sensors monitor important performance indicators like load stress, hydraulic pressure, vibration frequency, and engine temperature.
- AI models are able to instantly start corrective steps when they identify anomalous patterns, such as a loader vibrating outside of typical operating ranges or a conveyor motor exhibiting symptoms of overheating.
    - To avoid mishaps, machines might be isolated, rerouted, or shut down.
    - Accurate diagnostic data is sent to maintenance personnel, enabling focused repairs prior to a

malfunction turning into a dangerous situation.

- This automated method guarantees that machinery contributes to safe, continuous operations, minimizes error, and lessens the need for human supervision.

6.8.4. Integrated Hazard Intelligence

- AI-driven danger identification creates a comprehensive safety picture by combining operational, structural, and environmental data.
- By connecting several data streams, such as miner position, machine performance, vibrations, and gas levels, artificial intelligence (AI) may identify intricate, compound hazards that conventional systems could overlook.
- **Example:** AI can identify the combined danger and initiate preventive safety procedures if a water intrusion event occurs concurrently with structural stress in a nearby tunnel and an overloaded loader is operating nearby.
- As a result, a proactive, multi-layered safety system is created in which risks are recognized, ranked, and eliminated immediately—long before they become emergencies.

In summary, underground mining safety is transformed from reactive observation to proactive intelligence through AI-driven danger detection. Predictive analytics, automatic equipment checks, and real-time monitoring build a network of constant vigilance in which each piece of data helps to reduce risk. Mines in this AI-enhanced environment are able to predict dangers and react more quickly, protecting equipment, people, and business continuity.

6.9. Next-Gen AI Hazard Detection System

By 2035, AI-powered underground mining hazard detection systems will develop into *completely self-sufficient, self-learning safety ecosystems* that can anticipate, stop, and eliminate hazards without continual human assistance. Predictive analytics, equipment intelligence, and environmental monitoring are all combined into one intelligent network by these next-generation technologies.

6.9.1. Autonomous Multi-Sensor Systems

- Mines will equip machinery, shafts, and tunnels with dense networks of intelligent IoT sensors.
- In order to provide AI engines with real-time data, these sensors will continuously record environmental parameters including temperature, humidity, vibration, seismic activity, and gas levels.
- By correlating data from several sensors, advanced AI will spot minute patterns that are invisible to the human eye, such early-stage gas build up or micro fractures indicating an imminent roof collapse.

6.9.2. Predictive and Self-Learning AI Models

- AI models will be able to anticipate the emergence of threats with previously unheard-of accuracy, in addition to detecting anomalies.
- AI will predict the following using operational patterns, environmental models, and historical data: o Structural failures before stress exceeds critical thresholds.
- Equipment breaks down before dangerous vibrations or

temperatures occur.

- Incidents of gas or water intrusion before levels or inputs become dangerous.
- In order to improve future hazard projections, these systems continuously learn from each incident and modify predictive algorithms in response to results.

6.9.3. Adaptive Risk Assessment and Reaction

- The system will automatically rank hazards according to their operational impact, proximity to personnel, and severity.
- Individual miners will receive individualized real-time notifications that will guide them along the safest pathways and offer precautionary advice.
- Autonomous control of machinery and infrastructure will eliminate the need for human intervention by rerouting or shutting down machinery, deploying emergency barriers where necessary, and adjusting ventilation to remove hazardous gasses.

6.9.4. Humans and Machines Working Together Autonomously

- As AI oversees real-time safety, miners will shift from being reactive responders to strategic overseers, concentrating on making decisions.
- Continuous situational awareness will be provided via wearables, augmented reality (AR) gadgets, and smart helmets that show risk zones, safe paths, and hazard maps.
- Robotic support units and autonomous rescue drones will be prepared to respond immediately, reducing human exposure to risk by neutralizing hazards, delivering supplies, or directing evacuation.

6.9.5. Result: An Ecosystem of Self-Regulating Safety

- AI continuously monitors, anticipates, and reduces risks in the mine, turning it into a live safety network.
- The system can handle several threats at once and continue to function normally, resulting in incidents being reduced to almost nil.
- Strategic planning replaces crisis response in human oversight, guaranteeing the smooth coexistence of operational resilience, productivity, and safety.

In conclusion, the next-generation AI danger detection system sees underground mines as self-governing, intelligent ecosystems where robots serve as both tools and protectors, personnel are proactively guided, and problems are foreseen. *By 2035, underground mining will be safer, smarter, and more resilient than ever thanks to this self-learning, fully integrated technology, which will also establish a new global benchmark for operational safety.*

6.10. Dynamic Evacuation Systems

Underground mining emergencies necessitate prompt, coordinated actions, regardless of the cause—gas leaks, floods, structural instability, or equipment breakdowns. Conventional evacuation plans may not work well in situations that change quickly because they rely on fixed pathways, human guidance, and generic alarms. Dynamic evacuation systems powered by AI transform emergency response by giving miners, cars, and rescue crews adaptive, real-time instructions:

6.10.1. Real-Time Route Optimization:

- AI continuously tracks tunnel integrity, airflow, water levels, and hazards throughout the mining environment.
- Routes are dynamically recalculated to guarantee that miners are never sent to locations that become unsafe during the evacuation process; algorithms determine the quickest and safest evacuation routes, taking into account blocked passages, collapsing portions, or areas with gas concentration or fire.
- AI can immediately simulate several evacuation scenarios through integration with mine maps and digital twins, selecting the path that reduces each miner's or group's exposure to risk.

6.10.2. Personalised Alert and Guidance

- Each miner receives customized notifications from wearable technology, such as smart helmets, AR glasses, or wristbands, which offer:
    - Directions to the closest secure exit, turn by turn.
    - Alerts regarding new dangers that may appear in their route.

    - Updates on assembly locations and evacuation schedules.
- If new dangers emerge or if congestion forms in specific tunnels, these instructions automatically reroute miners in real-time. In order to guarantee that miners in high-risk areas receive instructions first, the system can also prioritize notifications for them.

6.10.3. Self-sufficient Assistance Cars

- Autonomous loaders, transport trucks, and AI-controlled drones offer vital assistance during evacuations:
    - **Guiding staff:** Especially in low-visibility situations, vehicles guide miners along safe routes.
    - **Transporting supplies**: Without human assistance, emergency kits, oxygen tanks, or medical supplies are given to people in need.
    - **Rescue of injured personnel:** Autonomous vehicles can transport miners who are incapable of moving independently, guaranteeing prompt removal from hazardous regions.
- These autonomous vehicles function in tandem with the AI network, allowing for smooth integration with miner sites and real-time hazard data.

6.10.4. Command and Control Integrated with AI

- Real-time tracking of miner movements, vehicle locations, and changing risks is done by centralized AI systems that keep an eye on all evacuation operations.
- In the event of unanticipated events, control centres can strategically step in by rerouting miners, deploying more autonomous support, or modifying AI priorities.

- By ensuring that evacuation decisions are based on comprehensive, precise, and real-time data, this integration lowers confusion and raises survival rates.

6.10.5. Advantages of Systems for Dynamic Evacuation

- **Quicker Evacuation**: AI shortens the time needed to direct miners away from peril.
- **Enhanced Safety:** Real-time hazard adaptation in routes and instructions reduces risk exposure.
- **Operational Resilience:** Rescue crews experience less human strain and avoid mistakes brought on by panic thanks to automated help.
- **Scalability:** Systems are able to coordinate thousands of personnel at once during large-scale evacuations.

Conclusion: Adaptive, AI-driven emergency management has replaced static, human-dependent processes in underground mine safety with dynamic evacuation systems. Mines establish a safety network that can effectively guide each person to safety, even in the most complicated and dangerous circumstances, by integrating autonomous support vehicles, individualized alarms, and real-time route optimization.

6.11. Response to Integrated Emergencies

Emergencies in underground mining can get out of hand in a matter of seconds. With numerous tunnels, large equipment, and hundreds of workers, modern mines are complicated and require a coordinated reaction that is beyond human capacity. AI-powered emergency response systems ensure that crucial choices and actions are carried out promptly and precisely by establishing a smooth interface between people, machines, and surface control centres.

6.11.1. Dashboards for Centralized Monitoring

- AI creates a consolidated dashboard by combining data from wearable technology, IoT sensors, machines, and environmental monitors.
- Complete visibility of subterranean conditions, such as temperature, vibrations, water intrusion, gas levels, structure integrity, and personnel locations, is sent to surface control centres.
- Rather of depending on fragmented information or delayed notifications, operators and emergency managers can see the whole picture in real-time thanks to this holistic view.
- Predictive overlays and interactive simulations are additional features of advanced dashboards that illustrate the emergence of possible hazards and the effects of different reaction tactics.

6.11.2. Automated Decision-Making and Prioritization

- AI continuously examines incoming data to pinpoint the most pressing dangers.
- The system ranks reactions according on:

    - Hazard severity (e.g., roof collapse versus slight gas fluctuation)
    - Personnel proximity to danger areas
    - Possible effects on operations
- Critical actions, including turning on alarms, rerouting ventilation, turning off dangerous machinery, or sending out autonomous rescue vehicles, are guaranteed to be taken right away thanks to automated decision-making.
- AI also supports human supervisors by offering suggested interventions supported by past event data and predictive analytics, which enhances decision-making accuracy and speed under duress.

6.11.3. Communication Networks That Are Dependable and Redundant

- Because of environmental variables, lengthy tunnels, and rock interference, underground mines are infamous for having communication problems. AI-powered emergency systems guarantee that instructions and alarms are continuously transmitted: o Wearables, smart helmets, and augmented reality devices are used to notify miners.
    - Autonomous cars and machines are given operational instructions instantly.
    - Surface control centres communicate with subterranean staff and equipment in both directions.
- Even in challenging or changing circumstances, failsafe communication is made possible by redundant networks, such as mesh radio, fibre optics, and Internet of Things connectivity.

6.11.4. Coordinated Human-Machine Response:

- Integrated emergency response technologies augment human capabilities rather than take the place of human operators.
    - AI works with human teams to coordinate autonomous cars, drones, and rescue robots.
    - AI autonomously controls ventilation, environmental controls, and hazard containment while real-time guidance is provided to miners to safe positions.
    - Rescue activities are coordinated to reduce exposure and increase effectiveness, guaranteeing that all available resources are used wisely.

6.11.5. Constant Feedback and Learning

- AI records and examines each incidence, near-miss, and response, generating ongoing learning loops:
    - New patterns are automatically reflected in emergency protocols.
    - AI finds weaknesses in reaction tactics and makes recommendations for enhancements.
    - As the system improves in speed, accuracy, and adaptability over time, the likelihood of reoccurring events decreases.

6.11.6. Benefits of Integrated Emergency Response

- **Quicker Reaction Times:** AI speeds up the process of identifying threats and launching countermeasures.
- **Reduced Human Risk:** Employee safety is maintained when high-risk tasks are handled by autonomous systems.
- **Operational Continuity**: During emergencies, damage and downtime are minimized by intelligent cooperation.
- **Improved Situational Awareness:** Real-time updates and

centralized dashboards guarantee that all stakeholders are operating with correct and comprehensive information.

In conclusion, the pinnacle of AI-driven safety in underground mining is integrated emergency response. Mines can create a proactive, coordinated, and adaptable safety network by integrating automated decision-making, centralized monitoring, dependable communication, and human-machine collaboration. Even in the most difficult subterranean circumstances, this technology maintains operational resilience, handles risks promptly, and guides personnel safely.

6.12. Case Study: Canadian Nickel Mine

An impressive illustration of how AI and IoT change underground mining safety from a reactive procedure to a proactive, predictive system can be found at a Canadian nickel mine. The mine is exposed to a number of hazards due to its deep underground operations, such as structural instability, flooding, and gas leaks. The incorporation of a next-generation AI-driven safety ecosystem was necessary because traditional safety measures by themselves were unable to ensure timely, coordinated reactions.

6.12.1. Continuous Monitoring using IoT Sensors

- IoT sensors were placed throughout the mine's shafts and tunnels to continuously monitor the following:
    - **Air quality:** amounts of oxygen, carbon monoxide, and methane.
    - **Structural stress** includes tunnel strain, vibration patterns, and rock stability.
    - **Equipment health**: temperature, vibration abnormalities, and load stress.
- AI algorithms evaluate data streams from thousands of sensors in real-time, separating normal variations from potential dangers.

6.12.2. Modelling Predictive Risk

- AI models use sensor readings, geological surveys, and historical data to predict possible risks like:
    - Water intrusion from adjacent aquifers or old tunnels.
    - Roof instability brought on by tension build-up.

    - Equipment malfunctions that can result in operational risks.
- By using these predictive models, mine operators can foresee issues well in advance of their escalation, allowing miners and support systems to take preventive measures.

6.12.3. Incident Reaction: Event of Water Ingress

The efficacy of the technology was proved by a small water intrusion incident:

- **Early Detection:** Thirty minutes before conventional threshold-based sensors would have sent out alarms, the AI system identified rising water levels. The critical lead time to initiate emergency procedures was given by this early notice.
- **Automated Mitigation:** Autonomous pumps were set up right away to securely reroute water, avoiding flooding in locations that were vital and safeguarding equipment.
- **Coordinated Evacuation:** Wearable technology provided miners in impacted locations with real-time alerts, directing them along prearranged evacuation routes. As circumstances changed, AI dynamically modified the directions to make sure the safest routes were taken.
- **Result:** Proactive hazard management proved to be effective as the incident was contained without causing any injuries or interruptions to operations.

6.12.4. Lessons Learned:

This example highlights a number of important benefits of AI-powered safety systems:

- **Proactive Safety**: AI reduces risk before events happen by anticipating dangers and changing the mine's strategy from reactive alarms to preventive intervention.

- **Smooth Human-Machine Coordination**: Miners may concentrate on safe evacuation while automated systems take care of risky jobs like rerouting machinery or pumping water.
- **Operational Continuity:** Activities go on without hiccups even in the event of an emergency, lessening the financial and logistical toll that incidents take.
- **Continuous Improvement**: AI models are fed data from every incident, increasing the effectiveness of hazard detection and evacuation in the future.

6.12.5. Broader Implications

The case of the Canadian nickel mine shows that integrating AI and IoT redefines mining safety culture and goes beyond disaster prevention. Miners work in a setting where risks are foreseen, controlled, and eliminated instantly, boosting self-assurance, output, and general resilience.

In conclusion, this mine's success serves as an example of how artificial intelligence (AI) has the potential to revolutionize underground mining. AI makes it possible to go from reactive safety procedures to a predictive, life-saving strategy by identifying hazards early, coordinating automated actions, and safely directing workers. This sets a global standard for contemporary mining operations.

6.13. The Future: Mines That Are Ultra-Safe and Self-Optimizing

The idea of safety is changing from a reactive requirement to a basic, integrated aspect of every operation as mining moves into the era of powerful AI and IoT. In order to reduce risk, save lives, and improve operational efficiency, tomorrow's mines will be extremely safe, self-optimizing ecosystems where predictive intelligence, autonomous intervention, and continuous learning come together.

6.13.1. Self-Learning, Cognitive Mines

- Mines of the future will function as completely integrated cognitive networks that continuously analyse data from personnel, equipment, and environmental sensors.
- The mine will be able to learn from experience and improve prediction models over time as a result of every incident, near-miss, and operational anomaly feeding into AI algorithms.
- In order to stop accidents before they happen, this learning process makes sure that risks are identified sooner, operational strategies are continuously improved, and hazards are assessed more accurately.

6.13.2. Multi-Factor, Real-Time Hazard Prediction

- By combining data from:
    - Internet of Things sensors that track temperature, structural stress, and gas concentrations, AI will assess environmental, operational, and human aspects all at once.
    - Machine health systems that identify wear, failure, or unusual vibration.

- Wearable technology for miners that provides activity patterns, physiological status, and location AI will detect new dangers milliseconds before they manifest by evaluating these intricate data streams in real-time, allowing for proactive mitigation as opposed to reactive action.

6.13.3. Autonomous and Adaptive Emergency Response:

- Humans, robots, drones, and surface resources will all be coordinated instantly by an entirely autonomous and adaptive emergency management system.
- AI will:
    - Dynamically determine the best escape routes in the event of a dangerous situation.
    - Use driverless cars to carry or lead miners.
    - Automate the control of environmental systems, such as fire suppression, lighting, and ventilation.
- Even when humans are unable to act swiftly enough to preserve safety, these systems react rapidly to changing circumstances.

6.13.4. Industry-Wide Safety Intelligence:

- Hyper-safe mines won't function independently; AI networks will compile knowledge from several mines throughout the world to build industry-wide knowledge bases.
- Mines will implement best-in-class procedures by comparing safety procedures and danger trends, thereby raising operational standards over time.
- Accurate hazard predictions and emergency response will be improved by predictive models trained on data from various

sites, transforming each catastrophe into a chance for group development.

6.13.5. Embedded Intelligence: Safety by Design

- Safety will no longer be a secondary concern; rather, it will be incorporated into the design of subterranean systems:
    - Tunnels, ventilation, and machinery will all be automatically adjusted to reduce danger.
    - Unsafe operations will be actively prevented by autonomous equipment.
    - AI and IoT will guarantee dynamic intervention and ongoing monitoring throughout the mining process.
- This method makes protection an inherent part of the operational environment by removing the conventional divide between danger detection and reaction.

6.13.6. Advantages of Extremely Secure Mines

- **Near-Zero Incidents:** Workplace accidents are significantly decreased by ongoing prediction and intervention.
- **Operational Efficiency**: Since accidents no longer result in expensive downtime, safety and productivity are linked.
- **Human Empowerment**: With the aid of AI-guided insights, miners evolve from reactive responders to strategic decision-makers.
- **International Knowledge Exchange:** Industry-wide learning guarantees that safety advances benefit all mines, resulting in a safer and more intelligent mining sector.

In summary, efficiency, intelligence, and safety are all essential to underground mining's future. Mines that are hyper-safe and self-optimizing will predict dangers, react on their own, and keep becoming better through learning. Safety is transformed from a reactive necessity into a fundamental operational concept by integrating AI and IoT into the very fabric of mining operations, protecting human life while fostering unmatched productivity and resilience.

Chapter 7: Data as Currency: Operational Profitability, Predictive Maintenance, and Mining Analytics

Overview

Data is the new gold in underground mining, not a by-product. Sensor readings, machine telemetry, ventilation measures, geotechnical data, and worker movement logs—all of which were previously regarded as background noise—are now being acknowledged as strategic assets. Data must be gathered, examined, and turned into useful intelligence in the same way that ore is mined, refined, and processed into precious metals. This change is changing mining operations by presenting information as a currency that promotes more informed choices, increased productivity, and eventually increased profitability.

7.1. From By-product to Strategic Asset

Mining has historically been measured in tonnes of ore moved or ounces of metal produced, but today success is increasingly measured by how well a company manages, interprets, and leverages its data. Every environmental monitor installed in a shaft, every drill rig with sensors, and every haul truck equipped with GPS all become part of a vast digital ecosystem. The true competitive edge is not in the volume of raw data collected, but rather in how that data is mined for patterns, correlations, and predictive insights.

7.2. Predictive Maintenance – From Reactive to Proactive Operations

Predictive maintenance is one of the most effective uses of data in underground mining. Historically, equipment failures resulted in costly downtime, delayed production, and costly emergency repairs. AI systems can anticipate failures before they happen with the help of IoT-enabled machines that feed real-time data streams, such as vibration levels, engine temperatures, and hydraulic pressures.

Rather than responding to breakdowns, mine operators can proactively schedule maintenance when it's least disruptive, prolong the life of machinery, and optimize spare parts inventory. This not only lowers costs but also improves safety because fewer catastrophic breakdowns occur underground, leading to a smooth transition from firefighting to foresight.

7.3. Mining Analytics – Unlocking the Invincible Value

Advanced analytics, when combined with raw data, can identify inefficiencies that are invisible to the human eye:

- **Operational Optimisation:** Finding inefficient haul routes, duplicate cycles, or energy waste in ventilation systems is known as operational optimization.
- **Resource Recovery:** Real-time analysis of ore grade data to minimize dilution and increase extraction efficiency.
- **Safety Analytics:** identifying abnormalities in rock stability markers or gas concentrations before they become dangerous. In order to provide concise, useful recommendations, these analytics continuously analyse millions of data points, functioning as an unseen workforce. As a result, mining becomes a highly precise sector rather than one that is primarily reactive.

7.4. Operational Profitability: Converting Knowledge into Income

The bottom line shows the worth of data when it is handled as currency. Optimized fleet management improves fuel consumption and increases tire life. Better fragmentation results from more intelligent blasting techniques guided by analytics, which reduces downstream milling expenses.

Unplanned shutdowns that cause multimillion-dollar output losses are prevented with predictive maintenance. When applied to large-scale enterprises, even small efficiency gains add up to substantial increases in profitability.

Integration is crucial. Value is diminished by data silos, which keep information from the operations, maintenance, and geology departments separate. Cross-departmental insights are revealed when they are combined into a single digital platform: procurement can precisely predict components demand, ventilation systems can be dynamically modified based on equipment use, and maintenance schedules can be matched with production projections.

7.5. The Living Ecosystem of the Intelligent Mine

The intelligent mine is a new mining paradigm made possible by AI and IoT, which are more than just tools. Every machine, environment, and asset in this case is interconnected and continuously feeds a self-optimizing system.

Mines are now dynamic ecosystems that adjust in real time to geological, mechanical, and economic situations rather than being static activities. According to this concept, a mine can be both a producer and a processor, extracting minerals and turning data streams into useful information. The future of industry prosperity will be determined by the firms that are able to successfully mine both information and ore.

7.6. Data-Driven Mine

These days, underground mines are live digital entities rather than merely rock tubes. IoT networks, which are made up of thousands of connected devices, constantly keep an eye on every facet of the operation, producing a deluge of data that circulates through the mine's digital veins.

As the mine's *"Nervous System,"* digital twins—virtual copies of real assets and procedures—simulate situations, forecast results, and serve as a testing ground for enhancements prior to deployment.

As a result, there is a quantifiable digital footprint left by each action, movement, and environmental change in this data-rich ecosystem. This turns the mine into an open, quantifiable, and—above all—optimisable business.

7.7. Streams of Data Mining

7.7.1. Equipment Telemetry

All drill rigs, loaders, conveyor belts, and transport trucks are equipped with sensors. Metrics including engine load, vibration, hydraulic pressure, tire wear, temperature variations, and fuel consumption are all recorded by these devices. AI will eventually be able to recognize early warning indicators of harmful operating situations, inefficient fuel use, and component breakdowns. The mine starts to handle equipment as a renewable, well-maintained resource rather than as a disposable one.

7.7.2. Environmental Metrics

Conditions below the surface vary every second. IoT sensors track dust levels, humidity, structural stress in rock walls, and gas concentrations (CO, CH_4, and NO_2). AI systems are able to identify irregularities that are imperceptible to humans, such as gas leaks, ventilation bottlenecks, or roof collapses. Mines increase safety and cut down on wasteful energy use by continuously modifying airflows and checking stability.

7.7.3. Personnel Metrics

Wearable technology monitors subterranean workers' biometrics, location, and signs of weariness. This improves emergency response and guarantees adherence to safety procedures by enabling prompt alarms to be sent out in the event that a miner exhibits an elevated heart rate or is exposed to dangerous gasses. By matching human performance with operational requirements, AI can even improve shift patterns to lower accidents caused by weariness.

7.7.4. Production Metrics

Digital twins are used in production to simulate downtime, blasting outcomes, throughput rates, and ore quality. AI can detect inefficiencies in milling and crushing circuits or poor blasting techniques that promote dilution. Operations can minimize waste and enhance ore recovery with these insights.

7.8. From Overload to Optimization

The amount of data is enormous—terabytes every day, coming in from thousands of points—and without intelligent analysis, it becomes impossible to handle. Artificial intelligence (AI) steps in as the orchestrator here, as machine learning models sort through the noise and identify the signals that are important. AI replaces dashboards full of raw data with clear insights, predictions, and prescriptive actions, like:

- Changing ventilation to save energy while maintaining air quality;
- Sending out maintenance crews before a haul truck breaks down;
- Optimizing blasting sequences to improve ore fragmentation;
- Rescheduling shifts to lower fatigue-related risks.

7.9. The Self-Optimising Mine

The mine becomes self-optimizing when AI takes over as the decision-making engine. Data is now prescriptive and predictive rather than reactive. With the use of simulations and probability models, decisions that previously required weeks of analysis are now made instantly.

The self-optimizing mine:

- Adjusts dynamically to geological, mechanical, and environmental changes;
- Learns continually from each drilling, hauling, and processing cycle;
- Balances safety and profitability by coordinating production goals with the welfare of people and the environment.

In the end, this transforms the mine into a closed-loop intelligence system, a process that generates competitive advantage and continual development by not just extracting minerals but also deriving meaning from data.

7.10. A Day in the Life of a Fully Data-Driven Mine

At 6:00 a.m., in a fully autonomous, data-driven underground mine, operations are already underway—no human needs to press a switch.

- ***06:01 - Fleet Dispatch Optimization***

The mine's digital twin automates the assignment of autonomous trucks and loaders to the best haul routes, balancing production quotas with fuel efficiency, by running overnight simulations that take into account ore body data, haul road conditions, and demand forecasts. Each vehicle rolls out with a predictive maintenance check already verified.

- ***07:15 - Adjustments for Real-Time Ventilation***

Rising CO_2 levels in a freshly blasted stope are detected by IoT sensors. The AI-driven ventilation technology saves megawatts of energy by rerouting airflow just to the impacted areas rather than waiting for human intervention. This prevents the mine from being over-ventilated.

- ***09:30 - The Use of Predictive Maintenance***

The wheel motor of a transport truck exhibits strange vibration patterns. After identifying the abnormality and comparing it to past failure models, the AI plans a pre-emptive service stop for when the truck next experiences downtime. Because the algorithm predicted the need weeks in advance, the warehouse team has already staged the parts. There is no unscheduled downtime.

- ***11:45 - Worker Safety Alert***

Wearable technology for miners can identify increased heart rate and weariness. The worker is moved to a less demanding duty and a new crew member is sent to cover when the system initiates a shift reassignment. Rest and hydration are recommended via an AI-powered medical check, averting a possible incident before it occurs.

- ***14:00 - Adaptive Production Optimization***

The processing plant's ore grade sensors identify that the most recent batch of ore is of poorer quality than anticipated. The AI maintains recovery rates while using less energy by recalibrating the flotation and grinding parameters rather than continuing blindly. For increased precision, the technology records the disparity and feeds it back into geological models.

- ***16:20 – Predictive Hazard Detection:***

After seismic sensors detect minor rock stress shifts in a portion of the mine, AI models identify the risk of a rock burst within 12 hours, suspending work in the area, allowing automated equipment to continue production in unaffected areas, and providing engineers with accurate simulations of stress relief options.

By the time human managers log in from their offices, the AI system has already produced a profitability dashboard that shows fuel savings from optimized routes, energy savings from smart ventilation, cost avoidance from predictive maintenance, and productivity gains from adaptive ore processing. The mine's financial health is transparent and updated to the second.

7.11. The Future Reality

This is the course of mining, not science fiction. The mine transforms into a self-sustaining ecosystem by combining digital twins, IoT, and AI into a single intelligence layer. Humans shift from making the same decisions over and over again to strategic oversight, concentrating on sustainability, long-term planning, and innovation.

Profitability, safety, and environmental stewardship are all concurrent results of intelligent, automated decision-making in the data-driven mine of the future, rather than trade-offs.

7.12. Predictive Maintenance: Increasing Lifespan, Reducing Downtime

In the high-stakes environment of underground mining, equipment is both the heartbeat of production and one of the main cost centres. Millions of dollars in lost production might result from a single haul truck breakdown deep underground that stops operations for hours or even days. The two extremes of maintenance strategies have historically been scheduled maintenance, where parts are changed according to time intervals regardless of their actual state, and reactive maintenance, where action is only taken when a breakdown occurs. Both strategies are costly and ineffective; the former results in resource waste and early component replacements, while the latter causes costly downtime.

Predictive maintenance powered by AI offers a third route that is accurate, data-driven, and revolutionary from an economic standpoint. Predictive maintenance guarantees that machines run longer, safer, and more efficiently by continuously assessing the condition of the equipment, identifying abnormalities before they become failures, and initiating focused interventions.

7.12.1. Constant Monitoring—The Mine's Eyes and Ears

IoT-enabled machinery is equipped with a network of sensors that continuously track vital health indicators:

- **Mechanical:** hydraulic pressures, bearing wear, and vibration levels.
- **Thermal:** friction heat, cooling effectiveness, and engine temperatures.
- **Operational:** cycle counts, idle time, and fuel use.
- **Structural:** tire integrity, fatigue cracks, and frame stress levels.

Every piece of equipment has a living health record thanks to this continuous flow of telemetry. The technology offers a 24/7 diagnostic check in place of manual inspections, making sure that nothing is overlooked.

7.12.2. Anomaly Detection: Prompt Alerts Prior to Disaster

In order to differentiate between tiny deviations and typical operational patterns, AI models examine sensor data. An imbalanced shaft might be the cause of a vibration spike; injector wear could be suggested by unusual fuel consumption; and an imminent pump failure could be indicated by rising hydraulic temperatures. AI is not affected by data overload or fatigue like human operators are. It is capable of simultaneously scanning thousands of data and identifying correlations that are not evident to the human eye. This implies that problems are not only identified but also anticipated, frequently occurring days or weeks before they materialize.

7.12.3. Proactive Intervention: Only Make Repairs When Necessary

Predictive maintenance allows for surgical accuracy in interventions rather than waiting for a breakdown or overhauling equipment on set periods. Only when indicators indicate that maintenance is required—neither sooner nor later—will it performed.

For instance: If sensors reveal no wear, a drill rig that is scheduled for monthly bearing replacement may continue to operate for three months; on the other hand, a haul truck axle exhibiting stress anomalies may be serviced days before a potentially disastrous breakdown.

This method guarantees optimal machine availability and usage while also cutting down on needless labour and parts expenses.

7.13. The Effect on Profitability

Predictive maintenance has advantages for the whole mining operation:

- **Fewer Unplanned Stoppages**: Production plans are maintained, preventing equipment failures from causing a chain reaction of delays.
- **Reduced Repair Costs**: By identifying minor problems early on, costly overhauls are avoided before they become severe failures.
- **Extended Equipment Lifespan**: By keeping assets in top working order, return on investment is maximized.
- **Better Spare Parts Management**: Inventory management systems help forecast demand, which lowers capital invested in surplus parts.
- **Increased Safety**: By preventing unexpected malfunctions in subterranean environments, the chance of mishaps and dangerous circumstances is decreased.

For large-scale mines, a mere 5–10% improvement in equipment availability might result in millions of dollars more output per year.

7.14. From Cost Centre to Value Generator

Maintenance is no longer a necessary investment but rather a strategic driver of income thanks to predictive maintenance. Maintenance is viewed as a facilitator of safety, resilience, and productivity rather than as downtime or a sunk expense.

Future mines will view equipment failure as preventable rather than inevitable, with every repair opportunity anticipated, scheduled, and maximized.

7.15. Case Study: Rio Tinto's Autonomous Fleets

One of the biggest mining corporations in the world, Rio Tinto, invented predictive maintenance for its fleets of self-driving trucks and drills at the Pilbara iron ore mines in Australia. Hundreds of sensors on each truck record information on fuel usage, tire pressure, braking effectiveness, and engine performance. A centralized AI platform receives this data, and machine learning algorithms use it to look for trends throughout the fleet.

The outcomes are remarkable:

- Because possible faults are addressed before they happen, unplanned downtime has decreased by about 30%.
- The business states that engines and hydraulic systems have component lifespans that are far longer than those of conventional maintenance cycles.
- Rio Tinto can lower its spare parts inventory while maintaining availability when needed thanks to predictive analytics that are also included into procurement.

In one of the most challenging mining environments on the planet, Rio Tinto has maintained production consistency while saving millions of dollars a year by switching from reactive and scheduled maintenance to predictive, condition-based methods.

7.16. Operational Analytics: From Understanding to Implementation

Data collection is the first, not the last, step in contemporary underground mining. How well data is translated into action determines its actual value. Mines embrace a world of predictive and prescriptive intelligence with AI-powered data, moving beyond manual decision-making and hindsight reporting. Each environmental metric, equipment signal, and sensor reading goes into a decision-making engine that converts numerical data into operational plans.

The gap between understanding what is occurring and making decisions about it is filled by operational analytics. A mine that operates more financially, efficiently, and intelligently is the end outcome.

7.16.1. Resource Optimization: Removing Obstacles and Equilibrium Workloads

A loader waiting too long for a truck, a drill crew delayed by ventilation settings, or transport routes jammed during peak shifts are just a few examples of the inefficiencies that plague traditional operations.

AI analytics solves these issues by:

- Constantly analysing operational data to identify bottlenecks instantly.
- Reallocating manpower and equipment on a dynamic basis to make sure resources are where they are most needed.
- Optimizing haulage schedule to minimize downtime and maximize the use of trucks and loaders.

For instance, AI can reroute extra resources to balance output throughout the mine if one tunnel is operating below capacity because of equipment availability. This keeps production from slowing down and guarantees that every piece of equipment contributes to the highest possible throughput.

7.16.2. Energy Efficiency: Mining Wisely

Use Less Energy is one of the biggest expenses in underground mining, especially when it comes to running heavy equipment and ventilation. Energy management is transformed from a crude tool into a precise one using operational analytics:

- **Ventilation on Demand (VoD):** AI dynamically modifies airflow to only give fresh air where machinery or employees are working. This reduces wasteful energy use without sacrificing security.
- **Lighting Optimization**: Depending on activity levels and sensor input, smart systems can adjust the amount of lighting below ground.
- **Equipment Load Balancing**: AI divides up tasks to keep machinery running within ideal ranges for power and fuel economy, preventing misuse of specific components.

In order to achieve sustainability and profitability objectives, mines reduce their carbon footprint and operational costs by converting energy into a managed variable rather than a fixed expense.

7.16.3. Optimizing Throughput with Adaptive Production Timetables

Production timetables are set weeks or months in advance in traditional mining. In actuality, however, things alter hourly: environmental factors vary, machinery malfunctions, and ore quality varies. This dynamic reality is too much for static planning to handle.

Adaptive scheduling is introduced using AI analytics:

- Ore quality sensors help processing facilities decide whether to modify their blending or milling plans in order to optimize recovery.
- Real-time schedule reshaping based on equipment availability data ensures that a single region of unavailability doesn't disrupt the entire production cycle.
- AI-driven modifications are triggered by environmental circumstances, such as rising gas levels or unstable ground, and reroute staff and machines to safer, more productive places.

Even in the face of variable conditions, the mine maintains a stable output thanks to this ongoing recalibration. Operations proactively pivot in real time rather than responding to setbacks.

7.16.4. Closing the Loop: From Understanding to Action

The ability of operational analytics to bridge the gap between insight and execution is what gives it its power. Knowing that there is a bottleneck or that energy is being squandered is no longer sufficient; AI makes sure that quick action is done:

- Recommendations stem from insights.

- Suggestions turn into automated judgments.
- Decisions are carried out quickly and precisely, resulting in observable operational changes such as equipment reassignments, modified timetables, and recalibrated ventilation.

As a result of this change, the mine becomes a sentient, responsive organism that is always learning and adapting.

7.16.5. The Bottom Line

It is evident how operational analytics have an impact:

- Smarter operations that rely less on reactive fixes and more on predictive insights.
- Quicker decision-making: AI can direct changes in a matter of seconds rather than weeks.
- Increased productivity, attained by making better use of already-existing resources rather than by acquiring additional equipment or labour.

Operational analytics driven by AI essentially reveal the hidden value in each hour of machine time, each litre of fuel used, and each ton of ore extracted. It serves as a link between unprocessed data and actual profitability.

7.17. Safety Meets Profitability

For a long time, safety in underground mining has been presented as a moral and legal requirement—a cost of conducting business rather than a source of income. However, this viewpoint is out of date in the age of artificial intelligence and data-driven operations. Profitability and safety are now closely related and no longer conflicting concerns. Every accident prevented saves millions of dollars in lost productivity, downtime, legal risk, and reputational harm in addition to saving lives.

Real-time monitoring, predictive maintenance, and AI-driven danger detection demonstrate that a safer mine also yields higher profits.

7.17.1. Avoiding Expensive Downtime

Shutdowns are frequently caused by accidents, equipment malfunctions, and dangerous situations, and they have an impact on the whole value chain. A collapsed tunnel or a fire in a piece of machinery causes long-term bottlenecks, disrupts logistics, and lowers output in addition to stopping operations in one place.

AI lowers this danger by:

- Identifying mechanical irregularities, ground stress, or gas accumulations before they become more serious.
- Planning maintenance to prevent disastrous subterranean malfunctions.
- Redirecting resources to other active regions while automatically stopping operations in dangerous locations.

Continuous production with fewer unplanned pauses is the end result.

7.17.2. Reducing the Cost of Compensation and Insurance Premiums

Risk exposure is used by insurance firms to determine premiums. Strong safety records are rewarded with favourable conditions, while mines with a high frequency of incidents are penalized with higher prices. Businesses show lower operational risk by implementing predictive safety systems, which results in:

- Lower insurance premiums.
- A decrease in claims for workers' compensation.
- Fewer penalties from the law and regulations.

What once depleted budgets turns into a quantifiable source of profit.

7.17.3. Safeguarding Human Resources - The Real Asset

People cannot be replaced, but machines can. The most precious and irreplaceable resource in a mine is its skilled labour force. In addition to causing personal sorrow, accidents also result in a loss of talent, bad morale, and diminished confidence in leadership.

AI-powered safety systems make sure that:

- Employees are constantly checked for physical stress, exhaustion, and exposure to dangerous gasses.
- Real-time notifications direct evacuations and medical interventions, enabling prompt emergency responses.
- A culture of safety is strengthened, making workers feel protected and appreciated.

Preserving human capital improves productivity, retention, and the organization's reputation, all of which contribute to increased profitability.

7.17.4. The Productivity-Safety System The flywheel

The self-reinforcing cycle that AI produces is its most potent safety impact:

- Less downtime results from fewer accidents.
- Increased productivity results from less downtime.
- Increased productivity leads to increased profitability.
- Increased investment in safety innovation results from higher profitability.

Efficiency and safety reinforce one another to create a flywheel that maximizes profits, with financial success being directly fuelled by the protection of people.

7.17.5. Mining Safety's Future

Safety systems will not only respond to risks but also foresee them as mines become increasingly automated and digitally connected. Every improvement in safety will result in an increase in profitability thanks to real-time analytics that continuously balance production demands with worker well-being.

The triple bottom line—productivity, profitability, and human life protection—will determine the future of mining.

7.17.6. Case Study: The "Elimination of Fatalities" Program of Anglo American

With impressive results, the multinational mining behemoth Anglo American has incorporated AI and IoT into its safety strategy. The corporation implemented digital worker wearables, predictive analytics for equipment breakdowns, and real-time environmental sensors throughout its operations as part of its "Elimination of Fatalities" campaign:

- AI models initiate automatic ventilation modifications and pre-emptive evacuations when they identify dangerous gas levels and ground stability threats early.
- Wearables track employee biometrics and notify managers of stress and exhaustion before problems arise.
- Why Predictive maintenance guarantees fewer unplanned equipment failures that can result in accidents or necessitate production stops.
- **The result has been noteworthy**: Anglo American acknowledged digital safety technologies as a key component of its operational resilience and reported a dramatic drop in recordable events.

In addition to saving lives, the business also saved money on insurance and decreased downtime, demonstrating that safety expenditures yield quantifiable financial gains.

Chapter 8: Implementing Hybrid AI Architectures Underground: Edge and Cloud

Overview

Intelligence, dependability, and speed are critical in underground mining. When it comes to responding to emergencies, monitoring dangerous situations, or controlling autonomous machines, every second counts. Due to latency and connectivity issues, traditional centralized computing systems—which depend on distant servers—frequently perform poorly in underground settings, leaving operations open to delays that could jeopardize efficiency and safety. Hybrid AI architectures, which combine cloud intelligence above ground with edge computing below, hold the answer. Mining operations are made faster, safer, and much more effective by fusing strategic intelligence in the cloud with real-time response at the edge.

8.1. Comprehending Edge Computing in Mining

Edge computing places processing power in close proximity to the data source, where mining operations take place. Wearable technology, driverless cars, and Internet of Things sensors all process data locally, allowing for quick decision-making without the need for remote servers. These edge devices' AI algorithms evaluate data in real time, enabling crucial functions to take place in milliseconds, such machine shutdowns, hazard alarms, or autonomous car navigation.

Among the main benefits of edge computing underground are:

8.1.1. Real-Time Decision Making

Safety and efficiency are improved by prompt reactions to operating bottlenecks, environmental dangers, and equipment irregularities. Previously taking minutes, decisions now happen instantly.

8.1.2. Less Reliance on Bandwidth

In order to free up network bandwidth and guarantee that vital contacts continue unhindered, not all data must be sent to surface servers.

8.1.3. Increased Reliability

Local processing guarantees uninterrupted operations even in the event that surface connectivity is momentarily lost as a result of network interference or structural impediments.

8.1.4. Applications in the Real World:

- Real-time transport route optimization, obstacle avoidance, and tunnel navigation are all accomplished by autonomous haulage vehicles. After using edge-enabled autonomous loaders, haulage efficiency increased by 20% at one copper mine in Chile.
- Response times are reduced from minutes to seconds thanks to gas and environmental monitoring, which enables prompt ventilation or evacuation modifications.
- Worker safety wearables offer immediate alarms in the event that employees come into contact with hazardous circumstances, ensuring protection even when surface systems are not operational; predictive equipment maintenance tracks temperature and vibrations to prevent failures, decreasing unscheduled downtime by up to 35%.

Underground mines become intelligent, self-aware ecosystems thanks to edge computing, where humans, robots, and sensors work together to make choices in real time.

8.2. Cloud Computing: Scalable Intelligence

Cloud computing offers scale and strategic insight, while edge computing offers speed and local reaction. By combining data from several locations, the cloud makes enterprise-wide optimization, predictive modelling, and sophisticated analytics possible.

Cloud computing's primary mining capabilities include:

8.2.1. Advanced Analytical

The cloud finds operational patterns, performance trends, and inefficiencies across sites by analysing massive amounts of historical and real-time data.

8.2.2. Forecasting Models

Cloud-trained machine learning models are able to anticipate production results, improve maintenance schedules, and predict equipment breakdowns. One mine used cloud-driven predictive maintenance to cut unscheduled downtime by 30%.

8.2.3. Reporting and Collaboration

Surface engineers and executives may coordinate actions, keep an eye on subterranean operations, and make tactical and strategic data-driven choices thanks to centralized dashboards.

8.2.4. Optimization of Cross-Site

Enterprise-level coordination is made possible by cloud analytics, which balances energy allocation, equipment utilization, and production across several mining locations.

Together, the edge and cloud enable fast local decision-making while the cloud guarantees long-term optimization and strategic oversight, providing both immediate and thorough operational intelligence.

8.3. Hybrid AI Architecture - The Best of Both Worlds

Cloud and edge computing are easily integrated via hybrid AI systems, which capitalize on their respective advantages.

8.3.1. Local AI on the Boundaries

Manages low-latency activities like emergency warnings, machine coordination, and autonomous car navigation. These choices are made in real time beneath, guaranteeing operational effectiveness and safety.

8.3.2. Analytics for Cloud AI

Analyses massive information to find long-term patterns, improve maintenance plans, and guide business strategy. Cloud AI guarantees that mines are running strategically as well as efficiently.

8.3.3. Learning in Both Directions

Cloud-trained AI models are installed on edge devices, and as data is continuously gathered at the edge, cloud models are updated, gradually increasing adaptive intelligence and prediction accuracy.

Strategic oversight makes use of the whole breadth of AI analytics, while underground operations react immediately to shifting conditions thanks to this synergy, which produces flexible, adaptive, and intelligent operations.

8.3.4. Case Study: Chile's Autonomous Copper Mine

The effectiveness of hybrid AI architectures is demonstrated in a copper mine in Chile. The mine went into action:

- **Edge Nodes:** Sensors, drills, and autonomous loaders processed local data to make decisions in real time.
- **Cloud Platform**: To facilitate predictive modelling, operational, maintenance, and safety data from many locations were integrated.
- **Results:** Predictive maintenance and quick local responses resulted in a 33% reduction in equipment downtime.
- Throughput was improved by coordinated autonomous machinery, resulting in a 25% increase in operational efficiency.
- As a result of environmental monitoring and predictive hazard alerts, safety incidents dropped by 30%.

Diagram: Hybrid AI Architecture in Mining

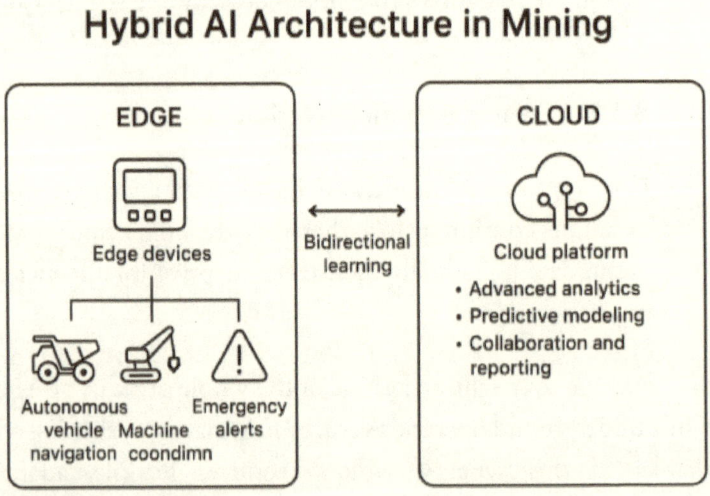

Figure 8: Hybrid AI Architecture in Mining

This example sets a standard for contemporary mining operations by demonstrating how real-time subterranean intelligence and strategic cloud analytics can turn a mine into a responsive, data-driven ecosystem.

8.4. The Future: Predictive, Cognitive, and Connected Mines

In the direction of fully cognitive subterranean systems, where mines function independently, effectively, and safely, hybrid AI architectures are developing:

- Without direct human intervention, mines will self-optimize, striking a balance between productivity, energy consumption, and safety.
- Edge AI will dynamically modify lighting, ventilation, and equipment operation, predict risks, and reroute autonomous cars.
- Cloud AI will benchmark performance, manage several locations, and guide business strategy.

Underground mining is now safer, quicker, and more lucrative than ever thanks to the edge-cloud synergy. Mines will become digitally sophisticated ecosystems rather than merely places for extraction, where:

- Cloud systems, edge devices, and sensors all fluidly exchange data; decisions are made instantly and predictively; and
- AI-driven insight directs every action.

Under this hybrid paradigm, mining turns into a cognitive, networked, and predictive activity in which AI, people, and machines work together constantly. The intelligence contained in every cubic meter of rock now defines the mining frontier rather than just depth or scale.

Chapter 9: Quantum Networks, 6G Integration, and Hyper-Intelligent Mines Overview

The conventional communication infrastructure is becoming outdated as mines become increasingly automated and networked spaces. Modern wireless, mesh, and edge-cloud networks allow for some coordination between sensors, equipment, and workers, but they are naturally limited by latency, interference, signal attenuation, and capacity. These difficulties are exacerbated by underground mining's intricate tunnels, metal obstacles, and heavy machinery.

By combining hyper-intelligent data orchestration, 6G connection, and quantum networks, the next phase of mining communication will overcome these constraints. In addition to faster and more dependable communication, these technologies offer a paradigm shift in mine operations, making them safer, more effective, and more predictive.

9.1. Security and Speed in Quantum Networks Reformulated

The concepts of quantum entanglement and superposition are used in quantum communication to send data with almost perfect security. This means that sensitive geological information and operational directives are protected in underground mines by:

- *Unhackable data linkages* between surface operations and subsurface systems.
- *Scalable sensor integration* that enables communication between hundreds of IoT devices without overloading the network;
- *Instantaneous coordination* between autonomous vehicles and robotic mining equipment, minimizing latency in crucial decision-making.

Quantum key distribution (QKD) can ensure that even in the case of interception, the data remains safe, providing miners with a level of cyber-resilience that traditional networks cannot equal.

9.2. 6G Integration: Pervasive Coverage and Ultra-Low Latency

Underground mining demands even greater performance, even though 5G has transformed industrial activities at the surface level. With sub-millisecond latency and terabit-per-second speeds anticipated, 6G networks will allow for the following:

- Real-time AI decision-making for self-driving vehicles, drills, and ventilation systems.
- Aggregation of vast amounts of sensor data, which feeds predictive analytics engines that can foresee geological dangers or equipment breakdown before they happen.
- Remote operators may monitor and operate mining equipment in immersive settings using high-fidelity AR/VR interfaces.

By overcoming the physical barriers of tunnels, advanced 6G technologies—like terahertz frequency bands and intelligent reflecting surfaces—can guarantee flawless connectivity deep beneath.

9.3. Hyper-Intelligent Mines: Predictive Autonomy and Edge-Cloud Convergence

A new class of hyper-intelligent mines is made possible by the combination of 6G connection and quantum-secured data channels, in which autonomous machinery, cloud AI, and edge computing devices function as part of a tightly coordinated ecosystem. Among its primary features are:

- Dynamic network adaptation, which modifies routing and communication bandwidth in real-time according to operational priorities and external circumstances.
- Workflows that optimize themselves, with AI managing ventilation, energy use, and mineral extraction without human involvement.
- Predictive hazard mitigation, which foretells collapses, gas leaks, or equipment failures in advance using simulations boosted by quantum technology.

Diagram: Hyper-Intelligent Mines

Figure 9: Hyper-Intelligent Mines

Underground mines will become cognitive ecosystems that continuously learn, adapt, and optimize themselves as a result of this technological convergence, replacing reactive and compartmentalized activities.

9.4. The Path Ahead

The use of 6G and quantum networks in mining is a strategic change rather than merely a technical advancement. Mines will develop into extremely effective, durable, and intelligent systems that can function in harsh environments with the least amount of danger to people. Therefore, the foundation of the next generation of mining, where speed, security, and intelligence come together to create genuinely hyper-intelligent operations, is underground communication.

9.5. A Revolution in Paradigms: Quantum Communication

Data transmission has undergone a radical change thanks to quantum communication, which offers capabilities that are much above those of traditional networking methods. It allows for extremely safe, quick, and intelligent information sharing that has the potential to completely transform mining operations by utilizing the concepts of quantum entanglement and superposition.

9.5.1. Extremely Safe Networks

Given the sensitivity of production indicators, geological data, and operational data in modern mining, security is of the utmost importance. This problem is addressed by quantum communication using Quantum Key Distribution (QKD):

Because of the nature of quantum states, any attempt at eavesdropping is instantly detectable, guaranteeing that operational data cannot be altered or stolen. For mines operating in remote or geopolitically sensitive regions, this ensures the integrity of communications between surface control centres and underground networks. QKD makes it possible to exchange cryptographic keys in a way that is fundamentally immune to interception.

9.5.2. Real-time Data Sharing

Near-instantaneous communication is possible with quantum entanglement, which could revolutionize subterranean mining operations:

- Real-time coordination between autonomous car fleets, drilling equipment, and conveyor systems can significantly increase productivity.
- Operators and safety systems can get critical alerts—like gas leaks, cave-ins, or equipment malfunctions—instantaneously, improving worker protection.
- Without latency delays, it is possible to navigate metal-dense surroundings and intricate tunnel networks, which typically weaken traditional wireless signals, guaranteeing continuous operational control.

9.5.3. AI Integration

Combining AI-driven network management with quantum communication unlocks its full potential:

- AI systems are able to continuously monitor quantum channels, anticipating interference or signal deterioration before it interferes with operations.
- Real-time data routing adjustments optimize information flow speed and dependability.
- To improve autonomous decision-making, operational planning, and predictive maintenance across mining systems, machine learning models can examine communication patterns.

9.5.4. Consequences for Mining Activities

Quantum communication has the potential to completely transform underground mining in the real world:

- **Fleet Coordination**: By operating in perfect sync, autonomous machines and vehicles can increase throughput and decrease downtime.

- **Real-Time Safety Alerts**: Quick danger identification and reaction systems reduce hazards and enhance worker safety.
- **Data Security and Compliance:** Company assets and intellectual property can be safeguarded by transmitting sensitive exploration and production data with complete assurance of confidentiality and integrity.
- **Improved Predictive Ability**: Mines can foresee equipment failure, geological instability, or energy shortages before they happen by sending safe, fast data into AI systems.

Essentially, quantum communication is a paradigm leap rather than just an improvement over conventional networks, setting the stage for mines that are not only quicker and more effective but also essentially safer, smarter, and more resilient.

9.6. 6G: The Upcoming Development in Underground Communications

Traditional wireless networks struggle to meet the speed, scalability, and reliability requirements of contemporary operations as mines grow more automated and data-driven. Although 5G has already revolutionized industrial communication at the surface level, 6G is the next big thing, providing previously unheard-of performance, intelligence, and flexibility for underground mining settings.

9.6.1. Speeds at the Terabit Level

Terabit-per-second data transfer speeds are anticipated from 6G networks, allowing for the real-time transport of enormous volumes of data:

- **High-Resolution Sensor Feeds:** Constant streaming from thousands of Internet of Things sensors that keep an eye on ambient conditions, structure integrity, and equipment.
- **Drones and Autonomous Vehicles:** smooth data transfer for telemetry, navigation, and coordinated fleet management in underground settings.
- **Advanced Imaging and Mapping:** 3D scans, LiDAR data, and geological models are instantly transferred for use in planning and decision-making.

9.6.2. Exceptionally Low Latency

Latency will decrease to almost zero, which is essential for both operational effectiveness and security:

- Robotic systems and autonomous equipment can react instantly to shifting circumstances.

- Without noticeable lag, remote operators can operate machinery from surface control centres or even off-site locations.
- Real-time AI decision-making makes it feasible for workflow improvement, hazard reduction, and predictive maintenance to happen instantly.

9.6.3. Slicing Networks to Increase Operational Effectiveness

Network slicing, a concept introduced by 6G, enables the development of specialized virtual networks suited to particular operational tasks:

- **Safety Networks**: Communication for emergency systems, environmental sensors, and alarms is given priority.
- **Telemetry networks** are high-throughput data channels used for predictive analytics and equipment performance monitoring.
- **Autonomous Control Networks**: Exceptionally dependable, low-latency channels for robotic operations and fleet coordination. This division guarantees that competing network traffic will never impede or delay vital operations.

9.6.4. Astute Networking

It is anticipated that 6G networks would self-optimize by utilizing AI to dynamically adjust to subterranean conditions:

- Continuous adjustments are made to signal strength, bandwidth distribution, and data routing in response to operational priorities, equipment density, and environmental changes.

- By anticipating communication bottlenecks and rerouting data in advance, predictive algorithms can reduce downtime.
- By bringing decision-making closer to the source, integration with edge computing lessens dependency on centralized cloud services and improves responsiveness even further.

9.6.5. Establishing Hyper-Networked Mines

Truly hyper-connected mining environments will be made possible by 6G when paired with IoT devices, edge AI, and autonomous systems:

- AI-powered equipment and human operators work together in real time with ease.
- Prescriptive and predictive insights are made possible by continuous, high-fidelity data streams, which improve operational safety and efficiency.
- Mines develop into robust, flexible ecosystems where every choice, from danger avoidance to energy management, is guided by real-time intelligence.

In conclusion, 6G is a revolutionary technology that will revolutionize underground mining communication, opening the door to new levels of intelligence, safety, and efficiency while establishing the groundwork for completely autonomous, ultra-intelligent mines. It is not just a minor improvement.

9.7. An Overview of Hyper-Intelligent Mines for 2035 and Later

The mining industry is expected to undergo a significant shift in the upcoming ten years due to the convergence of 6G networks, AI, IoT, and quantum communication. In addition to automating tasks, these technologies will build cognitive, adaptive, and predictive mining environments that can self-manage and guarantee the highest levels of sustainability, efficiency, and safety. By 2035, the idea of a mine will change from being a group of machines and people to an intelligent ecosystem in which all the components work together in real time with unmatched accuracy.

9.7.1. Self-Optimising Ecosystems

Mines will develop into data-driven, self-governing systems:

- AI is able to make real-time operational modifications through ongoing monitoring of environmental conditions, energy consumption, and output indicators.
- Machine learning algorithms will detect inefficiencies and dynamically reorganize workflows, increasing throughput while reducing waste and energy consumption.
- Automatic optimization of ventilation, lighting, and energy distribution will be based on workforce distribution, equipment load, and air quality.

9.7.2. Management of Predictive Hazards

Artificial intelligence and quantum-enhanced sensing will raise safety to previously unheard-of levels:

- To identify micro-failures before they become more serious, sophisticated sensors will track the stress levels in rock formations, tunnels, and support systems.

- Real-time sensor data analysis will be used to forecast seismic occurrences, gas leaks, and flooding, allowing for proactive mitigation.
- In order to protect human workers and reduce production downtime, alerts and autonomous responses will be activated quickly.

9.7.3. **Fleet-Autonomous Coordination**

6G and quantum networks will allow mining equipment to connect instantly, facilitating highly coordinated operations:

- A smooth, self-organizing fleet of trucks, loaders, drills, and drones will maximize material handling and ore transportation.
- AI-powered scheduling will balance workloads, reduce downtime, and dynamically adjust to shifting geological conditions.
- Because remote operators will have real-time oversight, surface crews can strategically intervene when needed without interfering with autonomous operations.

9.7.4. **Human-Machine Cooperation**

In mining, human involvement will change from manual labour to working in tandem with intelligent systems:

- Wearable sensors and AR/VR interfaces will give miners real-time operational guidance, hazard alarms, and contextual insights.

- AI systems and miners will collaborate in a symbiotic partnership where machine accuracy and human intuition work together to get the best results.
- Immersion simulations driven by AI and fast connection will improve situational awareness, training, and decision support.

9.7.5. Integration of Global Data

The hyper-intelligent mines will function as a component of an interconnected worldwide network:

- Mining companies will be able to coordinate activities across continents as if they were a single, cohesive site thanks to cloud and quantum communication networks.
- Predictive analytics and data-driven benchmarking will optimize production plans, safety regulations, and resource allocation throughout the company.
- In addition to facilitating sustainable practices, real-time worldwide oversight will guarantee environmental compliance and conscientious resource management on a large scale.

9.7.6. In summary: Moving Toward a Future in Cognitive Mining

Hyper-intelligent mines will completely change the mining industry by 2035 and beyond. These ecosystems will combine human creativity and machine intelligence to attain previously unthinkable levels of safety, efficiency, and sustainability. They will be autonomous, predictive, and intimately integrated. Mines will become living, learning, and adaptable systems rather than isolated, reactive activities, establishing a new benchmark for industrial innovation.

9.8. Case Study: The Hyper-Connected Mine Concept

A number of progressive companies are experimenting with hybrid 6G-IoT networks to develop hyper-connected mining settings as the mining industry transitions to complete automation and digital intelligence. Even though many of these projects are still at the conceptual or pilot stage, they provide a window into the mining industry's future, where operational norms will be redefined by the convergence of real-time data, AI-driven decision-making, and ultra-secure communication:

9.8.1. Self-governing Activities and Instantaneous Coordination

Autonomous drillers, loaders, and ore transporters in this hypothetical mine are continuously communicating positional and operational data via fast 6G networks:

- A self-organizing fleet of vehicles and equipment dynamically modifies its routes and timetables to optimize efficiency.
- Quick reactions to unforeseen challenges, such equipment delays or geological anomalies, are made possible via real-time communication.
- Centralized AI ensures constant productivity by keeping an eye on fleet activity and redistributing resources as needed.

9.8.2. Environmental and Equipment Safety Predictive AI

Cutting-edge AI algorithms examine sensor data from environmental monitoring systems and all equipment:

- **Equipment Stress Prediction**: Before a drill or transporter fails, AI can identify unusual vibrations, heat patterns, or wear signs.
- **Forecasting Environmental Hazards**: The technology may automatically initiate safety procedures and provide workers with proactive warnings by analysing gas concentrations, humidity, seismic activity, and structural stress in real time.
- **Maintenance Scheduling**: By optimizing maintenance cycles, predictive algorithms lower unscheduled downtime and increase equipment longevity.

9.8.3. Wearable Technology for Human-Machine Cooperation AR

Wearable technology powered by AI gives miners situational awareness and contextual guidance:

- Overlays of augmented reality (AR) show performance indicators, safety alerts, and instructions right in the miner's line of sight.
- Personalized AI helpers help employees navigate intricate tasks, lowering mistakes and improving security.
- Without being physically present in high-risk areas, remote supervisors can communicate with miners online and offer real-time assistance.

9.8.4. Communication Secured by Quantum

Quantum encryption is used to secure all communications inside the mine:

- Fleet directives and operational data cannot be intercepted or altered thanks to quantum key distribution (QKD).
- Corporate cloud networks, surface control centres, and subterranean systems securely exchange sensitive geological and production data.
- Large-scale, autonomous mining operations require a robust communication backbone, which is established by integrating quantum security.

9.8.5. Anticipated Results and Strategic Advantages

It is anticipated that implementing these cutting-edge technologies will produce revolutionary outcomes:

- **Efficiency Gains**: Predictive AI, autonomous fleet coordination, and optimized resource allocation all boost total productivity.
- **Near-Zero Downtime**: Predictive maintenance and real-time monitoring significantly lower operational disruptions and equipment failures.
- **Unprecedented Safety:** Workers in formerly high-risk areas are protected by proactive danger identification, AR-guided human-machine collaboration, and automated emergency responses.
- **Data-Driven Decision Making:** Management is able to make strategic, well-informed decisions about production, logistics, and sustainability thanks to continuous, high-fidelity data streams.

This hypothetical hyper-connected mine serves as an example of how the mining industry can be drastically altered by combining 6G, IoT, AI, AR, and quantum communication. In this future scenario, mines function as safe, intelligent, and adaptable ecosystems that bridge the gap between human knowledge and mechanical accuracy.

9.9. The Hyper-Intelligible Mine Era

Communication in underground mining will undergo a radical change in the future rather than just a technical advancement. The combination of quantum networks, 6G connectivity, and hyper-intelligent systems heralds the beginning of a new era in which mines are no longer merely reactive industrial sites but rather cognitive, adaptive, and fully interconnected ecosystems, even though AI and IoT have already shown their ability to optimize operations, improve safety, and increase profitability.

1. **Secure and Instantaneous Communication**

6G networks and quantum communication will do away with latency, interference, and bandwidth restrictions:

- Information is instantly transferred between autonomous fleets, subterranean machinery, and surface control centres.
- Sensitive business, geological, and operational data is protected from cyberattacks and tampered with thanks to quantum encryption.
- Mines can function with previously unheard-of assurance regarding the accuracy and velocity of their transmissions.

1. **Proactive and Predictive Activities**

The paradigm of mining will change from reactive to predictive due to hyper-intelligent systems:

- AI algorithms that are fed data from high-resolution sensor networks predict environmental risks, workflow bottlenecks, and equipment breakdowns.

- Real-time predictive analytics, as opposed to post-event reactions, inform maintenance, resource allocation, and operational decisions.
- Mines develop into self-optimizing, adaptive ecosystems that constantly learn from their surroundings.

1. **Smooth Interaction Between Humans and Machines**

Humans will now work closely with intelligent systems rather than only observing them:

- Wearable AR devices and AI-driven guidance allow miners to engage with autonomous machinery safely and efficiently.
- By interpreting and combining human input with machine intelligence, cognitive systems establish a mutually reinforcing workflow in which accuracy and intuition support one another.
- Immersion, real-time information improves training, danger response, and operational planning by bridging the gap between AI insight and human expertise.

1. **Intelligent Ecosystems and Data-Driven Decision Making**

Data becomes the lifeblood of operations in hyper-intelligent mines:

- Every machine, sensor, and environmental parameter contributes to networks enabled by the cloud and quantum, generating an endless supply of useful information.
- Automated, accurate, and context-aware decision-making is used for everything from fleet routing to energy management, ventilation optimization, and safety procedures.
- The subterranean habitat itself develops into an intelligent,

responsive ecosystem that can modify its own parameters to preserve sustainability, safety, and efficiency.

1. **Toward a Future That Is Smarter, Safer, and More Connected**

AI, IoT, quantum networks, and 6G will all come together to reinvent mining in ways that were previously unthinkable:

- There will be less danger to human workers since mines will be safer.
- There will be less downtime and better use of resources due to more efficient operations.
- Intelligence and connectivity will turn subterranean spaces into dynamic industrial systems where people, machines, and data live side by side peacefully.

In summary, the next generation of mining involves a fundamental rethinking of what a mine may be, not just a change in technology. With the help of AI, IoT, and next-generation communication technologies that characterize the era of hyper-connectivity, hyper-intelligent mines will usher in a future in which underground operations are autonomous but human-centric, predictive yet adaptive, secure yet highly collaborative.

Chapter 10: Pilot Projects and Case Studies: Practical Applications Changing Mining Activities

Overview

Only through practical application does theory become transformative. In an effort to reinvent underground operations, mining corporations worldwide are progressively using AI, IoT, and other cutting-edge technology. These solutions are actively changing workflows, improving worker safety, and increasing operational profitability; they are not merely theoretical. Through in-depth case studies and experimental projects, this chapter examines the concrete effects of these advancements, demonstrating how technology is changing the parameters of what is feasible in contemporary mining.

10.1. The Rationale for Real-World Pilots

Because of the size, challenging conditions, and safety-critical nature of operations, integrating AI and IoT in mining is challenging. Through pilots, businesses may evaluate ideas in safe environments, gauge their efficacy, and improve their rollout plans. Validating predictive maintenance models for vital equipment is frequently one of the key goals:

- Improving underground monitoring and communication in real time.
- Reducing the impact on the environment and energy use.
- Increasing worker safety by proactively identifying hazards.

10.1.1. Case Study 1: Operational Efficiency and Predictive Maintenance in a Chilean Copper Mine

An AI-powered Internet of Things platform was used by a large copper mine in Chile to track the condition of its machinery and improve production schedules.

Important results included:

- **Reliability Of The Equipment**: 35% fewer failures occurred, saving millions on repairs.
- **Production Throughput**: Because loaders, crushers, and conveyors coordinated optimally, ore output rose by 20%.
- **Energy Efficiency:** AI-guided ventilation and intelligent load balancing reduced energy usage by 12%.
- **Workplace Safety:** AI-assisted hazard detection resulted in a 28% decrease in safety occurrences.
- **Lessons learned:** Sensor placement and data quality were crucial. Early interaction with operators reduced opposition to new technologies and ensured its seamless deployment.

10.1.2. Case Study 2: Edge AI and Autonomous Vehicles at the Australian Gold Mine

To improve subterranean logistics, a gold mining company in Australia used edge AI systems and driverless haul trucks. Results showed:

- **Smooth Navigation**: AI systems dynamically rerouted cars to steer clear of traffic jams and obstructions.
- **Less Downtime for Operations**: Predictive analytics found bottlenecks before they became serious.
- **Safety Improvements**: Automated car systems reduced the amount of time people spent in dangerous areas.

The significance of strong edge computing was demonstrated by this project, which allowed for real-time decision-making even in situations with poor connectivity.

10.1.3. Case Study 3: Integrated Safety Systems in a Platinum Mine in South Africa

For labour surveillance, a platinum mine in South Africa combined wearable technologies, IoT sensors, and artificial intelligence. Among the major accomplishments were:

- **Real-Time Location Tracking**: To stop workers from entering dangerous areas, their movements were tracked.
- **Environmental Monitoring:** Sensors automatically generated alerts based on temperature, air quality, and gas levels.
- **Predictive Health and Tiredness Monitoring:** Information from wearables helped lower workplace accidents.

This initiative's success emphasizes that in order to improve safety outcomes, technology adoption must give human-centric design first priority.

10.1.4. Knowledge Gained via Pilot Initiatives

Over various projects, a number of recurring themes surfaced:

- **Information Is Money:** AI performance is driven by extensive, high-quality datasets.
- **Computer-Human Cooperation:** Automation and operator knowledge are balanced in an effective application.
- **Interoperability Is Crucial:** Systems must smoothly interface with current mine management software. Phased pilots lower risk and promote buy-in.

Incremental adoption is the most effective approach.

10.1.5. Routes to the Hyper-Networked Mine

These practical applications show how to create fully hyper-connected mines:

- Underground real-time decision-making will be managed by Edge AI.
- Cross-site operational optimization will be made possible by cloud-based analytics.
- Hybrid AI systems will dynamically manage staff and equipment, anticipate risks, and optimize energy use.
- Workflows between humans and machines that are completely autonomous and cooperative will become commonplace.

Mining firms can expedite the transition from pilot to full-scale adoption by examining these achievements and difficulties, eventually converting underground operations into highly productive, robust, and intelligent ecosystems.

10.7. Key Takeaways from Various Case Studies

This chapter's case studies and pilot projects provide important insights into what makes modern mining operations successful. Even though every mine has a different operating environment, a number of universal lessons have been found that can direct the use of AI, IoT, and advanced analytics in the future.

10.7.1. Integration Is Essential

Isolated innovations rarely lead to success in modern mining. Integrating edge computing platforms, IoT devices, communication networks, and AI algorithms into a single ecosystem is necessary for true change. Real-time optimization is made possible when various systems cooperate, allowing data to move smoothly from sensors to operators to decision-support platforms. Conversely, inadequate integration may result in disjointed insights, underutilized machinery, and lost efficiency potential.

Example: Only when IoT sensor data was completely integrated with the equipment scheduling system did predictive maintenance analytics at a Chilean copper mine work well, enabling AI to generate proactive alerts that truly avoided downtime.

10.7.2. Productivity Is Driven by Safety

AI-powered safety systems improve operational efficiency in addition to saving lives. AI decreases accidents and unscheduled stoppages by tracking personnel movements, monitoring environmental conditions, and forecasting equipment breakdowns. Regulatory compliance, employee morale, and productivity all rise in mines that put safety first.

Example: Wearable technology and gas monitoring sensors cut safety occurrences in a platinum mine in South Africa by 28%. This improved throughput overall, prevented expensive shutdowns, and safeguarded workers.

10.7.3. Data Is Strategic

In isolation, raw data is not very useful. Streams of sensor data are converted into useful intelligence by predictive analytics and ongoing monitoring. By using this intelligence, mines can predict production bottlenecks before they happen, optimize ventilation and energy use, and foresee equipment problems. Essentially, data turns into a strategic tool that supports well-informed decision-making at every level.

Example: Edge AI systems at Australian gold mines maximized ore transport efficiency while minimizing downtime and congestion by dynamically rerouting vehicles based on real-time feedback from autonomous haul trucks.

10.7.4. Flexibility Is Important

Mining environments are dynamic by nature; infrastructure, operating priorities, and geological conditions can all change quickly. IoT devices, communication networks, and AI systems must be flexible enough to adjust to changing circumstances without requiring expensive reconfiguration or prolonged outage. Mines can adjust to unanticipated difficulties while preserving operating continuity thanks to flexible systems.

Example: Adaptive ventilation control greatly reduced energy usage while preserving air quality in Chilean copper operations by adjusting airflow according to ore extraction and machine load.

10.7.5. Cooperation Between Humans and Machines

Even though AI and autonomous systems are strong, the most prosperous mining operations depend on a mutually beneficial partnership between people and machines. Instead of trying to completely replace human decision-making, technology should support it. AI cannot completely replace the context, expertise, and intuition that skilled operators provide. Mines can accomplish goals that neither could accomplish on their own by fusing human understanding with machine accuracy.

Example: Human drivers worked alongside autonomous trucks in Australian gold mines, stepping in to handle complicated situations. Through this partnership, safety was guaranteed, efficiency was increased, and confidence in the self-governing systems was increased.

Diagram: Lessons Learned Across Case Studies

Figure 10: Lessons Learned Across Case Studies

10.8. Conclusion: Making the Dream a Reality

These case studies' lessons show that technology improvement is insufficient on its own. A comprehensive strategy that integrates systems, puts safety first, strategically uses data, adjusts to changing conditions, and promotes human-machine cooperation is essential for success. In addition to being more successful and productive, mines that follow these guidelines are also setting the stage for the next wave of intelligent, hyperconnected mining operations.

This chapter's case studies and pilot projects offer convincing proof that AI, IoT, and advanced analytics are already useful, deployable technologies that significantly enhance mining operations rather than just theoretical ideas. These technologies are actively changing underground mines all around the world, from autonomous coordination that maximizes output to AI-driven safety systems that save lives to predictive maintenance that lowers expensive equipment failures.

10.8.1. **From Awareness to Action**:

The power of AI and IoT to transform unprocessed data into useful intelligence is what gives them their true worth. Mines that use intelligent automation, predictive analytics, and continuous monitoring achieve operational foresight that was previously unthinkable.

Real-time insights from a network of machines, sensors, and cloud-based analytics platforms can now inform decisions that previously depended only on human judgment. This change minimizes operational expenses and environmental impact while allowing mining businesses to maximize output, decrease downtime, and improve worker safety.

10.8.2. **A Transformation Roadmap**:

For other mining businesses looking to undergo digital transformation, examining successful projects offers a clear road map. The following are important lessons learned:

- Give top priority to integrating AI, IoT, edge computing, and communications for smooth operations.
- Put an emphasis on safety-first strategies that also increase productivity.
- View data as a strategic resource that facilitates predictive decision-making for environmental, human, and equipment systems.
- Create systems with flexibility in mind, guaranteeing robustness in dynamic, constantly shifting subterranean environments.
- For best results, embrace human-machine cooperation and use operator knowledge in conjunction with autonomous systems.

10.8.3. **Looking Ahead:** The knowledge gained from these innovative projects serves as a guide for underground mining's upcoming phase. Operations in the future will be resilient, self-optimizing, and hyper-connected, with intelligent technology actively guiding and supporting decision-making. Mines will evolve from reactive to proactive ecosystems that can foresee problems, reallocate resources dynamically, and continuously enhance performance without the need for human intervention. Essentially, the technologies examined in these case studies signify a fundamental change in the way mining is carried out rather than only minor advancements.

Mining firms can fully realize the potential of intelligent, data-driven operations and build a safer, more effective, and more sustainable underground mining future by converting their vision into workable strategy and taking note of early accomplishments.

These case studies show that the groundwork for a new mining era has already been established. The achievements, difficulties, and knowledge gained offer concrete evidence that intelligent systems may convert subterranean activities from disjointed, reactive processes into completely integrated, predictive ecosystems. The following chapter expands on these observations by examining the idea of hyper-connected, self-optimizing mines in which autonomous systems, IoT, and AI coexist peacefully with human operators.

Here, we explore the operational approaches, architectures, and technologies that will shape underground mining's future and make the promise of digital transformation a reality.

Chapter 11: The Path Ahead: Constructing Hyper-Intelligent and Resilient Underground Mines

Overview

A straightforward reality emerges from the exploration of AI, IoT, and next-generation communication technologies: underground mining will have a connected, intelligent, and resilient future. Tomorrow's mines won't be passive places where both machines and people battle risk and inefficiencies. Rather, they will be self-optimizing ecosystems where AI anticipates risks, data informs decisions, and human-machine cooperation produces productivity never before possible.

11.1. Future-Ready Mine Principles

1. **The Foundation Is Connectivity**
2. The foundation of underground operations is intelligent communication networks that combine wireless mesh, edge computing, 5G/6G, and quantum technologies. Secure, low-latency, and seamless communication guarantees that all systems—human, robotic, and artificial intelligence—function in perfect synchronization.
3. **Data as Strategic Currency**
4. Every wearable, sensor, and self-governing equipment produces data, which artificial intelligence (AI) then turns into insights for hazard prevention, operational optimization, and predictive maintenance.

- Businesses that successfully use their data get a competitive edge while balancing efficiency and safety.
    1. **AI-Powered Intelligence and Autonomy**
- AI facilitates proactive, as opposed to reactive, decision-making in areas such as autonomous vehicle coordination and predicted hazard detection.
- As conditions change, mines may continuously learn, adapt, and self-optimize thanks to cognitive networks.
    1. **Human-Machine Symbiosis:**
    2. Technology should complement human knowledge rather than take its place.

- Miners who have wearable sensors, augmented reality interfaces, and real-time analytics become into strategic operators who work in unison with autonomous systems.
    1. **Design for Safety**
- Safety is transformed from a reactive necessity into an

inherent operating element through proactive danger reduction, automated evacuation, and predictive monitoring.

- Resilient mines protect productivity and life by anticipating hazards and reacting quickly.

11.2. Implementation Strategy Roadmap

11.2.1. Deployment of Digital Twins:

- Start by building a dynamic, real-time virtual model of subterranean operations.
- Combine operational, personnel, and environmental data to forecast and display system behaviour.

11.2.2. Internet of Things and Sensor Networks:

- Install dispersed sensors and actuators in wearable technology, machines, and tunnels.
- To sustain dependable data streams under changing subterranean conditions, make sure redundancy and flexibility are in place.

11.2.3. AI and Predictive Analytics:

- Use AI to detect hazards, optimize operations, and perform predictive maintenance.
- For better decision-making, continuously enhance models utilizing both historical and real-time data.

11.2.4. Edge and Cloud Hybrid Architecture:

- For both real-time reaction and strategic insight, integrate cloud-based analytics with low-latency edge computing.

11.2.5. Next-Generation Communication:

- For incredibly quick, safe, and robust communication, investigate 5G/6G and developing quantum networking.

- Construct communication networks with self-healing and adaptive routing powered by AI.

11.2.6. Pilot Projects and Continuous Learning:

- Measure results, scale gradually, and begin small with pilot implementations.
- Promote a culture of learning in which AI systems are continuously improved by operational insights.

11.3. A Prospect for the Upcoming Ten Years

Underground mines may function as completely hyper-intelligent ecosystems by 2035:

- *Humans and machines operating in perfect unison.*
- *Real-time operations optimization for profitability, efficiency, and safety.*
- *Systems for predicting hazards that stop mishaps before they happen.*

AI-driven insights are used to dynamically allocate energy, material, and human resources. Several sites, surface operations, and distant control centres are connected seamlessly.

In this future, mines will be more than just places to extract resources; they will be living, learning, and adapting spaces where AI and IoT will enable previously unheard-of levels of productivity, safety, and resilience.

Conclusion - The Deep Signal Imperative

The age of AI-powered, highly networked subterranean mining is already upon us. In a new era of smart, secure, and lucrative operations, businesses that adopt AI, IoT, edge and cloud computing, and next-generation communication networks will not only survive, but flourish.

Deep Signal is a call to action rather than just a technological roadmap. Mining firms can turn underground communication from a constraint into a competitive advantage and create the intelligent and resilient mines that the future requires by fully utilizing AI and IoT.

The deep signal is unmistakable: *underground mining has a bright, connected, and extremely efficient future, and it begins today.*

Table of Figures

Bibliography

1. *"Autonomous mining trucks and equipment – connected Pit Vipers are every miner's dream"*[1]*. 26 July 2018. Archived*[2] *from the original on 23 September 2020. Retrieved 10 February 2025.*

2. *"Caterpillar's autonomous vehicles may be used by NASA to mine the moon and build a lunar base"*[3]*. CNBC*[4]*. 23 October 2019. Archived*[5] *from the original on 25 October 2019. Retrieved 14 June 2025.*

3. *"Technical safety, control, lighting and monitoring system for mining mobile plant"*[6]*. www.mineserv.co.uk. Archived from the original*[7] *on 18 January 2016.*

4. *"Second landslide hits Rio's Bingham Canyon mine, 100 workers evacuated | Mining Australia"*[8]*. Archived from the original*[9] *on 27 September 2013. Retrieved 10 June 2025.*

---

1. *https://www.ericsson.com/en/blog/2018/7/autonomous-mining-trucks-and-equipment--connected-pit-vipers-are-every-miners-dream*
2. *https://web.archive.org/web/20200923105811/https:/www.ericsson.com/en/blog/2018/7/autonomous-mining-trucks-and-equipment--connected-pit-vipers-are-every-miners-dream*
3. *https://www.cnbc.com/2019/10/23/caterpillar-and-nasa-developing-autonomous-vehicles-to-mine-the-moon.html*
4. *https://en.wikipedia.org/wiki/CNBC*
5. *https://web.archive.org/web/20191025161343/https:/www.cnbc.com/2019/10/23/caterpillar-and-nasa-developing-autonomous-vehicles-to-mine-the-moon.html*
6. *https://web.archive.org/web/20160118025347/http:/mineserv.co.uk/mts-mining.html*
7. *http://www.mineserv.co.uk/mts-mining.html*
8. *https://web.archive.org/web/20130927014324/http:/www.miningaustralia.com.au/news/second-landslide-hits-rio-s-bingham-canyon-mine-10*

5. *"Bingham Canyon, Copper Mine, Utah, USA"*[10]. *Archived*[11] *from the original on 14 February 2021. Retrieved 10 June 2025.*

6. *"StackPath"*[12]. *17 April 2018. Archived*[13] *from the original on 11 August 2020. Retrieved 6 February 2021.*

7. *"How autonomous technology could make mining safer and more valuable"*[14]. *31 October 2019. Archived*[15] *from the original on 11 April 2021. Retrieved 4 June 2025.*

8. *"How automation is transforming mining's efficiency"*[16]. *27 April 2020. Archived*[17] *from the original on 10 June 2025. Retrieved 10 June 2025.*

9. *"The benefits and pitfalls of mining automation"*[18]. *Archived*[19] *from the original on 5 March 2021. Retrieved 6 February 2021.*

---

9. *http://www.miningaustralia.com.au/news/second-landslide-hits-rio-s-bingham-canyon-mine-10*

10. *https://www.mining-technology.com/projects/bingham/*

11. *https://web.archive.org/web/20210214051552/https:/www.mining-technology.com/projects/bingham/*

12. *https://www.smartindustry.com/blog/smart-industry-connect/the-normal-everyday-autonomous-mine/*

13. *https://web.archive.org/web/20200811153531/https:/www.smartindustry.com/blog/smart-industry-connect/the-normal-everyday-autonomous-mine/*

14. *https://www.nsenergybusiness.com/features/autonomous-mining-technology-zyfra/*

15. *https://web.archive.org/web/20210411184035/https:/www.nsenergybusiness.com/features/autonomous-mining-technology-zyfra/*

16. *https://www.raconteur.net/sustainability/automation-mining-efficiency/*

17. *https://web.archive.org/web/20210213174351/https:/www.raconteur.net/sustainability/automation-mining-efficiency/*

18. *https://www.mining.com/web/benefits-pitfalls-mining-automation/*

10. *"Press Release: Rio Tinto improves productivity through the world's largest fleet of owned and operated autonomous trucks"*[20]. *Archived from the original*[21] *on 29 April 2015. Retrieved 10 June 2025.*

11. *"Rio Tinto and Atlas Copco announce autonomous drilling alliance"*[22]. *Archived from the original*[23] *on 6 May 2017. Retrieved 10 June 2025.*

---

19. *https://web.archive.org/web/20210305010327/https:/www.mining.com/web/benefits-pitfalls-mining-automation/*
20. *https://web.archive.org/web/20150429095935/http:/www.riotinto.com/media/media-releases-237_10603.aspx*
21. *http://www.riotinto.com/media/media-releases-237_10603.aspx*
22. *https://web.archive.org/web/20170506045354/http:/www.riotinto.com/media/media-releases-237_1719.aspx*
23. *http://www.riotinto.com/media/media-releases-237_1719.aspx*

Don't miss out!

Visit the website below and you can sign up to receive emails whenever Patrick Mukosha publishes a new book. There's no charge and no obligation.

https://books2read.com/r/B-A-HJNZ-FLYZG

BOOKS 2 READ

Connecting independent readers to independent writers.

Did you love *Deep Signal: AI and IoT Unlocking Underground Mining Communication*? Then you should read *AI in Education: Teaching the Machines to Teach*[24] by Patrick Mukosha!

[25]

AI in Education: Teaching the Machines to Teach

What happens when machines not only assist in learning but begin to *teach*? This ground-breaking book takes readers inside the rapidly evolving world of Artificial Intelligence in education, where algorithms are reshaping how knowledge is delivered, assessed, and personalized.

From classrooms equipped with adaptive tutoring systems to universities experimenting with automated grading and virtual teaching assistants, *AI in Education* reveals both the opportunities and the challenges of this transformation. Rich with real-world examples,

24. https://books2read.com/u/mgO10q

25. https://books2read.com/u/mgO10q

it shows how AI can help close learning gaps, empower students with disabilities, and support teachers in managing ever-growing workloads.

But this is not just a story of technological advancement—it is a human story. The book explores the irreplaceable role of educators as mentors, motivators, and ethical guides in an age when machines can replicate many cognitive tasks but not empathy, creativity, or moral judgment. It raises critical questions about equity, ethics, and the future of work for teachers and learners alike.

Accessible, insightful, and forward-looking, ***AI in Education: Teaching the Machines to Teach*** is an essential read for educators, policymakers, technologists, and parents who want to understand—and help shape—the future of learning in an AI-powered world.

"Where human wisdom meets machine intelligence—the future of teaching begins." – Patrick Mukosha

Also by Patrick Mukosha

GoodMan

Resilient Strategies: Thriving in Harsh Business Conditions
Strategic Entrepreneurship: Navigating The Path To Success
Decisive Power: Navigating How to Make Toughest Decisions
"Reigning the Boardroom: A Trailblazing Guide to Corporate Governance Success"
Fortifying Digital Fortress: A Comprehensive Guide to Information Systems Security
"Unleashing the Power of Inclusive Innovation: Transforming the World for All"
"Exploring Computer Systems: From Fundamentals to Advanced Concepts"
"Computer Viruses Unveiled: Types, Trends and Mitigation Strategies"
"The Pinnacle of Success: Unveiling the World's 20 Most Successful Brands in 2023"
"Mastering Relational Databases: From Fundamentals to Advanced Concepts"
"Navigating Change: A Comprehensive Guide to Change Management"
"Information Systems Unraveled: Exploring the Core Concepts"
"Careers in Information Technology: Network Engineer"

“Careers in Information Technology: Network and Systems Administrator”
“Careers in Information Technology: Database Administrator”
"Careers in Information Technology: Cybersecurity Analyst"
“Careers in Information Technology: Cloud Security Specialist”
"Careers in Information Technology: Blockchain Developer"
"Careers in Information Technology: DevOps Engineer"
"Careers in Information Technology: Quality Assurance Analyst"
"Careers in Information Technology: Machine Learning Engineer"
"Careers in Information Technology: Artificial Intelligence (AI) Engineer"
"Careers in Information Technology: Internet of Things (IoT) Developer"
"Careers in Information Technology: AR/VR Developer"
“Careers in Information Technology: IoT Embedded Systems Designer”
“Careers in Information Technology: Artificial Intelligence (AI) Robotics Engineer”
“Careers in Information Technology: IoT Solutions Engineer”
“Careers in Information Technology: Data Scientist”
"Careers in Information Technology: Computer Vision Engineer"
"The Sociology of Information Technology"
“Smart Cities: The Technology Transforming Urban Living”
"The Future of Connectivity: Unleashing the Power of 5G and Beyond"
" Cybersecurity 2030 - Strategies for a Secure Digital Future"
The AI-Human Merger: Becoming Super-Intelligent Beings
AI Unleashed: The Future of Artificial Intelligence and Human Evolution
The Digital Prophet: Predicting the Next Decade of Disruption
AI in Agriculture: Farming the Future
AI in Education: Teaching the Machines to Teach

Deep Signal: AI and IoT Unlocking Underground Mining Communication

About the Publisher

Patrick Mukosha is an experienced ICT & Management Consultant. With 15+ years of IT and Management experience. He's passionate about all things IT & Management. He also loves to bring IT & Management issues down to a level that everyone can understand. His works have been widely quoted by Researchers and IT Practitioners. He has a PHD and MBA from AIU, USA; BSc(Hons) ICT, UEA, UK; Dipl, CCT, UK. He's a founder of PatWest Technologies.

www.ingramcontent.com/pod-product-compliance
Lightning Source LLC
LaVergne TN
LVHW090320210926
848480LV00001B/305
* 9 7 9 8 2 3 1 8 6 1 6 6 8 *